A PILOT CONSTRUCTED TREATMENT WETLAND FOR PULP AND
PAPER MILL WASTEWATER: PERFORMANCE, PROCESSES AND
IMPLICATIONS FOR THE NZOIA RIVER, KENYA

T0172539

Promoter: Prof. Dr. P. Denny
Professor of Wetland and Aquatic Ecology,
UNESCO-IHE Delft/Wageningen University,
The Netherlands

Co-promoter: Dr. J.J.A. van Bruggen,
Senior Lecturer,
UNESCO-IHE, Delft
The Netherlands

Members of the   Dr. T. Okia Okurut
Examination Committee   Lake Victoria Basin Commission
EAC Secretariat
Kisumu, Kenya

Prof. Dr. J. H. O'Keeffe
UNESCO-IHE, Delft
The Netherlands

Prof. Dr. J. T.A. Verhoeven
Utrecht University
The Netherlands

Professor Dr. M. Scheffer
Wageningen University
The Netherlands

A pilot constructed treatment wetland for pulp and paper mill wastewater: performance, processes and implications for the Nzoia River, Kenya

DISSERTATION

Submitted in fulfilment of the requirements of
the Academic Board of Wageningen University and
the Academic Board of the UNESCO-IHE Institute for Water Education
for the Degree of DOCTOR
to be defended in public
on Wednesday, 23 January 2008 at 15.30 hours
in Delft, The Netherlands

by

MARGARET AKINYI ABIRA

born in Nairobi, Kenya

Taylor & Francis is an imprint of the Taylor & Francis Group, an informa business

© 2007, Margaret A. Abira

Published by:
Taylor & Francis/Balkema
PO Box 447, 2300 AK Leiden, The Netherlands
e-mail: Pub.NL@tandf.co.uk
www.balkema.nl, www.taylorandfrancis.co.uk, www.crcpress.com

ISBN 978-0-415-46715-5 (Taylor & Francis Group)
ISBN 978-90-8504-815-2 (Wageningen University)

Dedication

To my husband and children for their patience and prayers.
Understanding is a fountain of life to those who have it. {Prov. 19:22}

Table of contents

Acknowledgements

This research study was undertaken as part of the Lake Victoria Environmental Management Programme (LVEMP, Phase I) activities to strengthen and improve the management of industrial effluents. The Ministry of Water and Irrigation (MWI) of the Government of Kenya implemented the Management of Water Quality and Ecosystems Dynamics Component of the Programme with funding from World Bank/IDA. I am indebted to MWI, for the scholarship that facilitated this study. I wish to thank Mr. Patrick Oloo, CEO of the Water Resources Management Authority for according me leave to undertake the thesis write up both at Delft and in Nairobi.

Prof. Patrick Denny's expert guidance as my Promoter was an immense privilege. I appreciate his sustained interest, commitment and constant encouragement throughout the study. My Co-promoter and mentor Dr. Hans van Bruggen was instrumental in shaping the vision and sustaining the work.

The study benefited greatly from the contribution of many people and institutions. I acknowledge the support of the LVEMP coordination office and the management and staff of the Water Quality Component. I am especially grateful to the Team members of the Integrated Tertiary Industrial Effluents Treatment Pilot sub-component of LVEMP. Peter Mwangi assisted with the design drawings and supervision of the wetland construction. Brainstorming sessions and assistance with field and laboratory work by Rose Ang'weya, Peterlis Opango, Charles Oleko, Charles Kanyugo, and Denis Owino, were a great help. Peter Muiruri assisted with microbiological examination while Henry Njuguna provided much guidance with statistical analysis. I appreciate the contribution of two M.Sc. students: Harrison Ngirigacha and Marlene Roberts. The data collection would not have been possible without the research field assistant, the indefatigable Mrs. Irene Belia Aketch and the cooperation of the management of Pan African Paper Mills (E. A. Ltd).

As part of the wider LVEMP project in Kenya I have over the past years come to appreciate and embraced linkages with various institutions and the scientific community in East Africa and elsewhere. I wish to acknowledge the role of the staff of the Kenya Marine and Fisheries Research Institute (Kisumu Centre); The School of Environmental Studies, and the Department of Technology (Civil and Structural Engineering Faculty) at Moi University; Jomo Kenyatta University of Agriculture and Technology, and the National Water and Sewerage Corporation of Uganda. The initial consultations with Prof. Frank Muthuri, (Kenya), Prof. D. Mashauri and Dr. Sixtus Kayombo (Tanzania), Dr. Tom Okurut (Uganda) and Dr. Karin Tonderski (Sweden) were stimulating.

The staff of UNESCO-IHE were always willing to assist in various capacities. I acknowledge the support and encouragement of staff in the Environment Resources Department, the Laboratory, ICT group, Library and Student Affairs. I cherish the moments and ideas I shared during the thesis write-up with fellow PhD participants especially, Rose Kaggwa, Julius Kipkemboi, Richard Buamah, Ruth, Sonko Kiwanuka, and Ashok Chapagain.

My inspiration partly came from my husband, Musa; sons Tony, Tom and Ted; and daughter, Nina. They believed that I could rise to the occasion the research posed and assisted by prayers, field sampling, data compilation, and thesis editing. I am grateful to God for your love. To Him be the glory forever.

During my stay in Delft, I was greatly blessed and encouraged through fellowship with the brethren of the Mount Zion parish of the Redeemed Christian Church of God, the members of the UNESCO-IHE Christian prayer group and the Mina hostel bible study group. May the Lord bless you always.

Margaret Akinyi Abira
May 2007

Chapter one

General introduction

General introduction

Kenya is an emerging country with a rapidly expanding population density and an eye to industrialization, agriculture and tourism to create wealth and alleviate poverty. As a result, the water resources increasingly are becoming polluted from both point and non-point sources. Municipalities and industries constitute the largest source of wastewater discharges. Industries discharge a variety of wastes some of which are toxic to human beings and the general environment. Examples of such industries, which are a critical environmental issue in Kenya today, include sugar, coffee pulping, textile factories, leather tanneries, pulp and paper mills, and slaughterhouses.

Lake Victoria basin

Lake Victoria in east Africa is the world's second largest freshwater body and is the source of the Nile River. The Lake has a water surface area of about 68,800 km^2 and is shared by Kenya (6 %), Tanzania (51 %) and Uganda (43 %). The Lake's catchment area is about 194,000 km^2 shared by Kenya (22 %), Tanzania (44 %), Uganda (16 %), Burundi (7 %) and Rwanda (11 %). The basin is home to about 34 million people providing them with water, food and various ecosystem goods and services.

Due to urbanization and unsustainable exploitation, the Lake Victoria basin's resources are currently facing a lot of threats and stresses, including land degradation, floods and droughts, fish stock depletion and water quality deterioration (LVEMP, 1996). Water pollution problems associated with industries are found throughout the basin, especially in the high potential land areas. The Lake Victoria basin in the western part of Kenya has not been spared. Several rivers whose catchments cover about 49,000 km^2 of high potential agricultural land with high population densities drain the Lake's catchment in Kenya. The total annual discharge is 7,292 million cubic metres and includes inadequately treated effluents from nine major municipalities and various industries such as distilleries, sugar, textile, leather and pulp and paper mill. This coupled with polluted runoff from agricultural and urban areas has had a profound effect on water quality and the ecology of rivers and the Lake (LVEMP, 1996). The result has been deterioration in water quality and environmental health, and a high cost of water supply since polluted water is expensive to treat. Water rights conflicts have increased due to shortage of water of suitable quality for domestic, livestock watering, industrial and irrigation purposes.

The sustainable management of the Lake Victoria basin environment and its natural resources, including water resources, requires an integrated approach with the participation of all stakeholders to achieve an effective reduction of pollutant levels; and to conserve its natural resources and enhance socio-economic development of the riparian communities. To this end, the Kenya Government, through the World Bank funding participated in the first phase of the regional Lake Victoria Environmental Management Project (LVEMP) initiated in 1994 through a Tripartite Agreement between Kenya, Uganda and Tanzania. The Management of the Water Quality and Land Use component of LVEMP has several sub-components including the Management of Industrial and Municipal Effluents. The Pan African Paper Mills (E.A.) Ltd (PANPAPER) at Webuye in western Kenya was identified as a site for an integrated tertiary effluent treatment pilot. This study was undertaken as part of the on-going LVEMP activities to strengthen and improve the management of industrial effluents. Sustainable management of industrial effluents calls for effective enforcement of environmental regulations by the regulating authority and adoption of cleaner production technology, as well as effective end-of-pipe treatment of effluents by industries.

The pulp and paper mill in Webuye discharges its treated effluents into the Nzoia River contributing the lion's share of all effluents discharged into the river with an estimated average discharge of 35,000 m^3 per day. This is undoubtedly a priority environmental issue as the final effluent quality seldom complies with government-prescribed effluent discharge guidelines. There is therefore need for a sustainable technology that can reliably achieve acceptable effluent quality for discharge into the environment at minimal cost.

Wetlands: definition, occurrence and values

Natural and artificial wetland systems have been used as a cost-effective alternative to conventional wastewater treatment methods for improving final effluent quality. Wetlands are "an ecotone- a transitional zone- between terrestrial and aquatic environments where water is the dominant factor in determining development of soils, fauna and flora. At least periodically, the water table is at or above the land surface" (Hammer and Bastian, 1989).

The Ramsar Convention on Wetlands of International Importance, especially as Habitats of Waterfowls of 1991, gives a broader definition. It describes wetlands as "areas of marsh, fen, peat land, or water, whether natural or artificial, permanent or temporary, fresh, brackish or salty, including areas of marine water the depth of which at low tide does not exceed six metres". This definition covers a wide range of marine, coastal, and inland habitats which (in Kenya) includes deltas, estuaries, mangroves and marine mud-flats as well as marshes, swamps, floodplains, shallow lakes and the edges of deep lakes and rivers containing vegetation influenced by light (Howard, 1992).

Hammer and Bastian (1989) argue that there is not a single correct definition of wetlands for all purposes. Most definitions seem to identify wetlands in terms of soil characteristics and the types of plants that the habitats support. Denny (1985), for example, favours the International Union of Conservation of Nature's (IUCN) definition which regards wetland areas as distinct from open water and characterised by having emergent and euhydrophyte vegetation.

Wetlands cover extensive areas (30 million km^2) of Africa (Thompson, 1985). They form about 13 % of the land area in Uganda (NEMA, 1996) while in Kenya they cover 2-3% (GOK/MENR, 1994). Kenya has a wide variety of wetlands. However, they are unevenly distributed with the majority located in higher rainfall areas in the central, western and coastal regions of the country. The low rainfall areas have few but important wetlands such as Lake Nakuru and the Shompole and Lorian swamps that are the home to a diversity of wildlife including flamingos and other birds, and provide expansive grazing areas for livestock and wildlife.

Wetlands have various functions and values which include: ground water recharge and discharge; water purification, storm protection and windbreak; shoreline stabilisation, biomass export, microclimate stabilisation, and fish nurseries. They are important sources of water for people, livestock and wildlife and provide food, building material, fuel-wood and medicinal products. Wetlands sustain habitats for unique bird and wildlife species and have cultural significance; for example, in parts of Kenya young initiates use wetland mud to smear their bodies during circumcision rights.

Wetlands provide free treatment for many types of water pollution (Hammer and Bastian, 1989). They can effectively remove or convert large quantities of pollutants from point sources (municipal and certain industrial wastewater) and non-point sources (mine, agricultural, and urban runoff) including organic matter, suspended solids, metals, and excess nutrients. Removal mechanisms for pollutants include natural filtration, sedimentation, biological decomposition and plant uptake.

Constructed wetlands

The use of natural wetlands for wastewater treatment is discouraged (Hammer and Bastian, 1989; Wetzel, 1993; Brix, 1993). This is due to, among other things, conflict of interest with various conservation efforts, the variable hydrology that is difficult to control, and difficulties in siting in relation to sources of wastewater. These and other factors have led to the rapid development of constructed wetlands in order to simulate and enhance the optimal properties of natural wetlands in performing the desired functions.

Constructed Wetlands are also known as "artificial wetlands" or "treatment wetlands" (Hammer and Bastian, 1989) and "reed bed treatment systems" (Cooper et al., 1996). Wetlands are engineered and constructed for four principal reasons as indicated by specific descriptive terminology (Kadlec and Knight, 1996): (1) to compensate for and help offset the rate of conversion of natural wetlands resulting from agriculture and urban development (constructed habitat wetlands); (2) to improve water quality (constructed treatment wetlands); (3) to provide flood control (constructed flood control wetlands); and (4) to be used for production of food and fibre (constructed agriculture wetlands).

Domestic wastewater treatment

Reed beds and other constructed wetlands for domestic wastewater treatment came to prominence in the mid-1980s in Europe (Cooper et al., 1996). The wetlands are usually constructed to provide secondary treatment of domestic sewage for village populations. In USA, Canada, and Australia, however, they tend to be aimed at tertiary treatment from towns and cities (Cooper et al., 1996). In sub-Sahara Africa, some studies have been carried out on treatment of domestic wastewater by natural wetlands in Kenya (Chale, 1985, 1987), and Uganda (Kansiime and Nalubega, 1999; Azza et al., 2000, and Kipkemboi et al., 2002). Studies using constructed wetlands have mainly been undertaken in South Africa (Wood and Hensman, 1989; Batchelor et al., 1990; Batchelor and Loots, 1997) but wider use in developing countries should be encouraged (Denny, 1997). More recently, Okurut (2000) carried out a study on secondary treatment of municipal wastewater in Uganda. The University of Dar es Salaam in Tanzania has also undertaken some studies (e.g. Mashauri et al., 2000; Senzia et al., 2002).

Industrial application

The use of constructed wetlands has spread to many other fields including industrial effluent treatment, acid mine drainage, agricultural effluent, landfill leachate and road runoff. Thut (1993) reports that by 1993 there were over 150 wetland systems in North America, ranging in size from a few sq. metres to over 500 ha. For the largest facilities, wastewater volumes of up to 80,000 m^3/day are being treated. These are comparable to the volumes generated by large pulp mills.

Several other facilities for industrial application are in the planning or early construction phases. These include installations of about 30–40 ha for the food processing (potato, meat and sugar), petroleum, and pulp and paper industries. There is therefore a lot of potential for constructed wetlands in the treatment of industrial effluents including pulp and paper mill effluents. However, the documented work on the performance of constructed wetlands for pulp and paper mill effluents has been carried out in the developed countries under temperate climatic conditions. It is not advisable unconditionally to translate their performance data and design guidelines for use in tropical Africa. Kadlec and Knight (1996) state that site-specific wastewater data showing historical flows and mass loads provide the best information for wetland system design. With regard to pulp and paper mill effluents, there are no published data for tropical climatic conditions. There is, therefore, need to undertake both laboratory and pilot scale studies to establish efficacy and optimal performance data for such constructed wetlands under tropical environmental conditions.

Literature Review

State of industrial wastewater management in Kenya

Agriculture is the main economic activity in Kenya, contributing to 30 % of the gross domestic product (GDP). The industrial sector currently contributes nearly 15 % of the GDP and is a growing source of exports (PricewaterhouseCoopers, 2006). The industrial sector is expected to grow further as Kenya strives to become newly industrialised by 2020. The major industries are those involved with processing of agricultural products such as coffee, sugarcane, cotton, hides and skins; and dairy products. In terms of environmental significance due to water abstraction and discharge of polluting effluents, especially to land and water, the main problem industries in Kenya are coffee, tannery, sugar, pulp and paper, slaughterhouses and dairy industries. Most of these industries discharge inadequately treated wastewater into the environment.

Environmental pollution is a growing concern in Kenya. That the government is conscious of this problem is manifested in the National Environment Action Plan (1994); the Policy on Environment and Development, and the Environmental Management and Co-ordination Act (2000), all of which express the need to protect the environment and ensure sustainable development. The government has already formulated guidelines and procedure for environmental impact assessment that requires all development projects, including industries, to provide measures for mitigating against pollution and environmental degradation. Further, clear policy guidelines have been laid down in the National Water Policy (1998) whose objectives include ensuring safe disposal of wastewater and environmental protection. The polluter pays principle will be applied. The policy states in part:

"...Levies on effluent discharge will be introduced based on the quantity of the effluent whose quality must conform to prescribed requirements of the standards in force..."

The Water Amendment Act (2002) which is the principal legislation governing the management of water resources addresses more effective control of water pollution. The National Water Resources Management Strategy for 2006-2008 (GOK/MWI, 2006) whose overall goal is to meet the water-related millennium development goals by 2015 outlines the approaches to pollution prevention while the recently gazetted Water Resources Management Rules (2007) specify regulations to be observed by all water resource users (including usage for waste discharge). The stage is, therefore, set for maximising the requirements for effluent quality. Industries will be required to comply with existing and future (likely) more stringent regulations aimed at reducing environmental degradation.

In the pulp and paper industry compliance with existing and future regulations can be achieved by adopting cleaner mill technology incorporating better internal process control measures to minimise pollutant loads, and external wastewater treatment technology that can achieve better final effluent quality. Cost-effectiveness and sustainability in terms of operation and maintenance are overriding factors in the choice of technology, especially for countries with developing economies like Kenya. Moreover, the pulp and paper industry globally is very capital intensive with small profit margins which tends to limit experimentation, development, and incorporation of new technologies into the mills (Akhtar and Young, 1998).

Pulp and paper manufacture

Paper manufacturing is divided into two phases: pulping the wood or other material and making the final paper product (Nemerow, 1978). The raw materials generally used in the pulping phase are wood, cotton or linen rags, straw, hemp, esparto, flax and jute or waste paper. These materials are reduced to fibrous pulp by either mechanical or chemical means. The process for making wood pulp begins with debarking. The bark is mechanically or hydraulically removed from wood before it is reduced to chips for cooking. Debarking is carried out either as a wet process or dry process, depending on the type of plant. To make mechanical (ground wood) pulp, wood is ground on large emery or sandstone wheels and then carried by water through screens. For chemical pulp wood chips are cooked or digested with soda (alkaline) sulphate (Kraft process), or acid sulphite (the sulphite process). The latter process results in weaker paper as compared to the former. The alkaline Kraft pulping process is by far the most important chemical pulping process practiced on an industrial scale. Pulps are refined, sometimes bleached (if a bright white product is desired) and dried. At the paper mill, which is often integrated in the same plant with the pulping process, the pulps are combined and loaded with fillers, finishers are added and the products transformed into sheets.

The processes at PANPAPER Mills
PANPAPER uses the Kraft pulping method. The following is a summary description of the pulp and paper production process at PANPAPER mills.

The main raw material is wood. Unbleached Kraft as well as bleached grades of paper is manufactured from plantation-grown softwood pines (*Pinus patula* and *Pinus radiata*) as well as cypress and some hardwood (eucalyptus). Pinus is harvested mainly from plantations in the Kaptagat forest about 80 km from the factory site. The average annual wood (chips) consumption for making chemical pulp is 271,000 m^3 pines, 27,000 m^3 eucalyptus, and 80,000 m^3 cypress. 90 % of the paper is made from chemical pulp while the rest is made from mechanical pulp. Mechanical pulp consumes only 9,500 tonnes of wood per annum, mainly cypress (70 %).

Debarked wood is ground with warm water to produce mechanical pulp. The water is not reused. The screened bark effluent contains fine particles of bark and wood and some dissolved solids. The resulting pulp is bleached using hydrogen peroxide (50 - 60 %) at a pH of 3 - 4 followed by hydrogen sulphite at pH 9 - 11. Sulphuric acid and sodium hydroxide are used to maintain the respective pH ranges.

To make chemical pulp, the chips are cooked or digested with cooking liquor. The liquor is produced from sodium sulphate, sodium carbonate, sodium sulphide and limestone. The average annual consumption of the chemicals is 270, 2800, 2400 and 3500 tonnes, respectively. The sulphate cooking process produces non-cellulose materials, lignins and resins, and mercaptans (methylsulphides, dimethylsulphides and dimethyldisuphides). The mercaptans, which are a source of odour, are condensed using water. The resulting black liquor is sent to the soda recovery plant for the recovery of alkali that is reused in the cooking process. The condensates, filtrates and cooling water are discharged into the effluent stream. The chemical pulp is screened, cleaned and then bleached with chlorine, 60 % as gas and 40 % as sodium hypochlorite. The pulp yield is about 50 % of wood.

At the paper mill, the refined unbleached or bleached pulp is mixed with chemicals such as soapstone powder, rosin/perschin (abietic acid), tamarind starch, dyes, guar gum, and alum among others. The average annual consumption is 31.8 tonnes rosin, 129 tonnes alum, 39 tonnes starch and 173 tonnes soapstone powder. The chemicals are used for sizing and other properties, which are desired in the paper. Once the right consistency has been achieved the pulp is screened on a belt of fine wire, the fourdrinier wire, pressed and dried on drying cylinders, rolled or calendared into desired shape before being rolled on reelers. The final product is cut into appropriate size sheets according to market requirements.

PANPAPER has a de-inking plant that uses the soap floatation method for recycled paper. The waste paper machine consumes 17,000 tonnes per year. When operating fully it accounts for about 2 % of the paper production. The factory also has an electrolytic plant with titanium and carbon electrodes where it manufactures chlorine, caustic soda and sodium hypochlorite used in bleaching pulp. The main raw materials are sodium chloride and sulphuric acid. The latter is used for drying chlorine gas. The plant makes hydrogen gas and hydrochloric acid as by-products.

Water required for the operations is abstracted from the Nzoia River. The factory has its own water purification plant. The freshwater consumption is 45,000 - 50,000 m^3 per day. The average effluent production is about 35,000 m^3 per day.

Characteristics of Kraft pulp and paper mill wastes

Pulp and paper industry wastewaters contain dissolved organic components of wood, including lignins, tannins, sugars, and cellulose; additives such as adhesives, sizing materials, starches, and resinates; and inorganic cooking chemicals (Gehm and Bregman, 1976). The main sources of waste at the pulp mills are the digester liquors (black liquors) while at the paper mill it is the screening and drying at the paper machines that produces "white water" (Nemerow, 1978). The characteristics of the effluents depend on the type of pulping and the internal measures employed by each industry to reduce pollution loading.

The discharge of dissolved substances from wet debarking is determined by wood species, the size of bark particles and contact time of the particles with water. The seven-day biochemical oxygen demand (BOD_7) load varies from 2 - 7 kg/m^3 wood (UNEP, 1985). The wastewater from softwood such as pines is usually highly toxic; 96h-LC_{50} values for fish are normally in the range of 2 - 3 % (UNEP, 1985). Effluent toxicity also arises from resin acids released during mechanical pulping. The BOD load from the dry debarking plant is about 0.1 - 0.5 kg/m^3 resulting from pre-treatment of logs with water or steam or from washing of the debarked logs.

In Kraft pulping, lignin forms methyl mercaptans, a source of odour in Kraft mills, and catechols (Elvers *et al.*, 1990). Catechols may subsequently be oxidized by air to quinones that largely account for the brown colour of Kraft pulp mill wastewater. If the mill uses chlorine to produce bleached pulp, some chlorinated organic compounds are produced. Some simpler compounds that have been identified include chloroform, chloroacetic acids, and various kinds of chlorinated phenols (Thut, 1993). Dissolved lignin material can also be chlorinated, producing large complex molecules which cannot be readily identified using even the most sophisticated analytical techniques (Thut, 1993). A general measure of chlorine attached to organic molecules is used to monitor this type of material. The most commonly used technique yields a measure of adsorbable (to activated carbon) organic halides (AOX). The chlorolignins are difficult to biodegrade. The white water from the paper mill contains inorganic substances that are added as mineral fillers, dyes and other additives such as starch, glue, brighteners, defoamers, slimicides, and dispersion and complexing agents.

Environmental importance of Kraft pulp and paper mill effluents

Besides the fact that the Kraft pulping process involves a very large capital investment, the major difficulties with the process is the accompanying obnoxious odour associated even with the most advanced mill, and the high water usage. Foam associated with Kraft mills is aesthetically undesirable. Foam is suspected to concentrate certain toxic components of effluent such as the resins and fatty acids (Water Pollution Control Directorate, 1976).

Untreated Kraft pulp and paper mill effluents can be toxic to fish. According to Howard and Walden (1974), although the effluents may be of low toxicity, the tremendous volume discharged from a Kraft mill "can produce deleterious effects in the environment to fish and other fauna that may be stressed by chronic exposure to sub-lethal effluent concentrations". Thus the organisms may be less able to meet natural stresses imposed upon them in the environment.

The Water Pollution Control Directorate (1976) and Johnston *et al* (1996) indicate that the toxicity of Kraft effluents is caused by, among other things, chlorinated lignins and phenols; resins and resin acids; slimicides, and sulphur compounds. However, the level of toxicity depends on other factors prevailing in the receiving stream such as pH, and dissolved oxygen.

Suspended solids are often associated with blanketing large areas of river bottoms causing a reduction in oxygen and the release of toxic hydrogen sulphide. The high load of suspended solids from pulp and paper mills, therefore, can destroy spawning grounds. Highly coloured effluents coupled with suspended solids reduce light penetration, thereby reducing primary productivity and altering the food web in the aquatic system (Water Pollution Control Directorate, 1976; Byrd et al., 1986).

Pulp and paper mill effluents also cause taste and odour problems in water and fish. The most apparent and widespread unpleasant effect noticed by the public is the taste and odour impairment (tainting) of fish flesh and fat (Water Pollution Control Directorate, 1976).

Treatment of pulp and paper mill wastewater

Pulp and paper mill wastes are treated in the following manner (Nemerow, 1978): (1) recovery (2) sedimentation and flotation to remove suspended matter, (3) chemical precipitation to remove colour, (4) activated sludge to remove oxygen-demanding matter, (5) lagooning, for purposes of storage, settling, equalisation, and sometimes for biological degradation of organic matter. However, at PANPAPER mills, treatment stages (3) and (4) are not followed. Nemerow (1978) reports that the cost of treatment is considered high in relation to the cost of product produced. Thus economic limitations have forced the industry to place emphasis on recovery rather than treatment. PANPAPER for example, recovers about 92 % of the lime used for making the cooking liquor.

Internal control measures
The environmental goal of internal process modifications is to reduce water consumption, reduce fibre losses and to reduce chemical and consequently BOD losses (Edde, 1984). These mainly involve recycling. The recovery processes in the paper mill involve the use of "save-alls", either in closed or partly closed systems. The save-alls are installed not only as a waste-treatment measure, but also as a conservation measure to recover fibres and fillers. Other measures include the following:

Chlorine bleach substitution: (i) Sequential chlorination – involves the substitution of a portion of the chlorine in the first bleach stage with chlorine dioxide. The preferred degree of substitution of chlorine dioxide for chlorine is 30 to 75 % on an equivalent basis. The equipment required to utilise sequential chlorination is only slightly different from that used in conventional chlorination, and it is relatively inexpensive to install (Edde, 1984). Sequential chlorination may be followed by a hypochlorite stage replacing conventional caustic extraction. This process change can result in a significant reduction in effluent colour (75 - 80 %) and toxicity, moderate reduction in effluent chlorides (20 - 25 %), chemical oxygen demand (COD; 20 - 35 %) and dissolved solids (20 - 30 %). PANPAPER Mills in the period 2002 - 2004 undertook a modernisation project with support from World Bank/IFC; aimed at achieving regional competitiveness, partly by replacing the chlorine bleach plant (World Bank, 1996). This has reduced the consumption of elemental chlorine by substituting it with chlorine dioxide. It is expected that the new process will improve the final effluent quality by reducing concentrations of absorbable organic halides (AOX) and dioxins. (ii) Oxygen bleaching – due to the high amount of chlorides usually present in the effluent from a conventional bleach plant, the incorporation of the bleach plant effluents, or part thereof, in the recovery system of a pulp mill has not been deemed technically feasible (Edde, 1984). The present state-of-the art indicates that elemental chlorine cannot be completely eliminated while producing fully bleached pulp, but the chlorine consumption can be drastically reduced by substituting part of the chlorine used with elemental oxygen as an oxidising agent. This achieves pollution load reductions of about 77 % for BOD_5, 87 % for colour, 76 % for COD and 70 % for chloride from the bleach plant effluent (Edde, 1984). Unfortunately, the oxygen reactor is more expensive than a conventional bleach tower (Edde, 1984). Thus such a modification in bleach technology may probably not be economically attractive for PANPAPER mills.

Closed-cycle Kraft mill: The closed cycle mill concept involves complete recycle of the unit process effluents, resulting in no process sewers and minimum purges. An example of a full scale closed cycle zero effluent mill is the Millar Western Meadow Lake Mill in Canada, which has been in operation since 1992 (Edde, 1994). The Mill's dissolved and suspended solids are concentrated and disposed by incineration. According to Edde (1994), closing the water cycle in a bleached Kraft mill will totally eliminate absorbable organic halides (AOX) and dioxin discharge problems that have become the major environmental issue in pulp mills since the 1990's. The closed cycle technology, therefore, has good environmental justification but in terms of commercial expenditure for an existing plant like PANPAPER it is not viable (Personal Communication with the Technical Manager of PANPAPER).

External control measures
The collected effluent is treated in the following stages (UNEP, 1985): (i) pre-treatment: In some mills bleach plant effluent is treated separately before mixing with the rest of the effluent stream for secondary treatment. (ii) Primary treatment: removal of suspended solids by sedimentation using clarifiers. (iii) Secondary treatment – removal of dissolved material by different biological methods viz.: stabilisation ponds (natural aeration, more than 10 days); mechanically aerated lagoons (5-10 days), activated sludge, trickling filter, anaerobic treatment.

In order to biologically treat pulp and paper mill wastes successfully some effluents must be neutralised or cooled first, and practically all must be supplemented with nitrogen and phosphorus since they are deficient in these microbial nutrients essential for high rate oxidation (Edde, 1984; Gehm and Bregman, 1976; Gapes *et al.*, 1999). The acceptable BOD_5:nitrogen:phosphorus ratio in a biological treatment system is usually considered to be 100:5:1 (Water Pollution Control Directorate, 1976). However, the actual nutrient requirement is a function of temperature and the growth phase of the microorganisms.

Wastewater treatment at PANPAPER Mills
The existing wastewater treatment system at PANPAPER Mills was designed, established and commissioned in 1974 handling 25,000 m^3/day of combined pulp and paper mill effluent. However, the discharged wastewater volume increased steadily with increased production and by the end of the 1990s had reached a maximum of 40,000 m^3/day with an average 35,000 m^3/day. The effluent is channelled through a bar screen into two parallel clarifiers then treated through a series of ponds two of which are aerated. A detailed description is given in Chapter 2. The effluent is finally discharged into the Nzoia River via a diffuser out-fall. Domestic wastewater separately goes to Webuye municipal sewage lagoons.

River Nzoia

The River Nzoia into which PANPAPER Mills discharges its wastewater is one of the largest rivers on the Kenyan side of the Lake Victoria basin. It has an average annual discharge of 52.0 m^3/s at Webuye (station 1DA02), the highest being 92.8 m^3/s in the month of September while the lowest flow is 10 m^3/s in March although flows as low as 0.2 m^3/s have been reported (JICA/GOK, 1992). The river flows through a densely populated area with 400-650 persons/km^2 (GOK, 1999) throughout the rest of its course before entering Lake Victoria. The river is important for artisanal fisheries and is a source of water to downstream riparian communities and industries such as the Mumias Sugar factory. In the dry season months of January to April, the dark colour of the effluent and foam is visible more than five kilometres downstream of the discharge point. Achoka (1998) reported the presence of partially-treated mill effluent in the river. Complaints of depleted fish stocks and presence of fibres in the water abound from the local communities who depend on the river for water supply.

Limited historic data available at the Ministry of Water Resources and Irrigation on the final effluent quality from PANPAPER shows that it seldom meets the discharge guidelines issued by the government of Kenya. There is, therefore, need to improve the final effluent quality for PANPAPER mills. This may be achieved by incorporating a constructed wetland for secondary or tertiary level treatment.

The study site

Pan African Paper Mills (E.A.) Limited commonly known as PANPAPER is the only integrated pulp and paper mill in Kenya. It is situated on the banks of Nzoia River in Webuye town, in the highlands of western Kenya. The mills are at global position 00°35'440"N 34°47'110"E and an altitude of 1640 m above mean sea level within the Lake Victoria basin, some 400 km northwest of Nairobi City (Figure 1.1).

The area has a tropical climate with mean air temperatures ranging from 14°C to 27°C. Two rainy seasons are experienced: the April-May "long" rains and the October-November "short" rains. The mean annual rainfall is 1671 mm (JICA/GOK, 1992). The relative humidity varies from 41-68 % and wind speeds are in the range of 1.0 to 1.7 m/s   (Meteorological data, Nzoia Station).

The pilot-scale constructed wetland at PANPAPER is located within the tree nursery below the final wastewater stabilization lagoon, about 1 km south east of the factory.

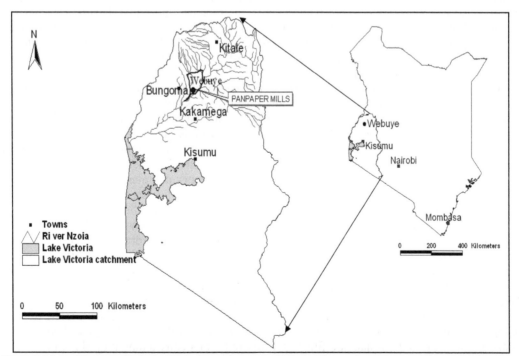

Fig. 1.1 Map of Kenya with Lake Victoria catchment and Nzoia River sub-catchment. The study site, Pan African Paper Mills, is located on the banks of Nzoia River in Webuye Municipality.

Research goal and objectives

The overall goal of the research study is to establish the potential of constructed wetlands for the treatment of pulp and paper mill effluents under tropical environmental conditions in order to improve the management of industrial effluents in Kenya, and in the East African Region in general.

The main objectives were to establish the efficacy of a pilot, experimental constructed wetland in the purification of pulp and paper mill wastewater, and to determine the optimal design criteria and operating conditions for a full-scale constructed wetland treating pulp and paper mill effluents. The specific objectives were:

- To characterize the effluent at PANPAPER mills.
- To establish the growth characteristics of various local emergent wetland plants such as *Typha domingensis*, *Phragmites mauritianus*, and *Cyperus papyrus* and *Cyperus immensus* in pre-treated pulp and paper mill wastewater.
- To determine hydrological and flow characteristics of the pilot constructed wetland system
- To establish the performance of three macrophyte species in removing nutrients, organic matter (BOD, COD), suspended solids, and phenols under varying operating conditions.

Thesis structure

The thesis focuses on exploring the use of a constructed wetland to improve the quality of the final effluent from PANPAPER Mills such that it is in compliance with national discharge regulations; and to protect the receiving aquatic environment and public health. The thesis is composed of seven chapters. The present chapter (Chapter 1) gives the background information on water pollution issues within the Lake Victoria basin and the Nzoia River, in relation to the pulp and paper industry. It also gives an overview of pulp and paper making processes and the produced wastewater treatment.

Chapter 2 provides information on the performance of the present wastewater treatment system of PANPAPER Mills. The nature and quantity of pollutants that are regulated by the government is detailed therein. This information forms the basis for the design of the pilot constructed wetland in Chapter 3. The chapter subsequently describes the key construction aspects, the set-up and operation conditions, hydrological (rainfall, evapotranspiration) and flow characteristics. Chapter 4 describes the experimental wetland system's establishment with respect to plants, these being the engineers of the systems treatment capacity. It includes information on biomass nutrient status and/or flows as an indicator of plant and ecosystem health. The nutrient removal capacity under different operation conditions and the implication for the water quality of the Nzoia River is discussed. The effect of plant growth coupled with various operation conditions (operation mode, hydraulic loading, hydraulic retention time) on treatment efficacy are discussed in subsequent chapters. Chapter 5 considers organic matter (BOD, COD) and suspended solids removal. Removal rates are estimated and implications for the water quality of the Nzoia River under various scenarios discussed. Chapter 6 addresses phenol removal. The initial findings in the first 15 months of operation (Abira *et al.*, 2005) are summarised in the chapter. The chapter details the removal processes and discusses the relative importance of various stages of plant growth on phenol removal and further gives budgets and reaction rates. In- and out-data are fitted to a first order model assuming plug flow. Finally, in Chapter 7 the main findings, conclusions and recommendations for design and operation are presented. The economics, and the future outlook including opportunities for full-scale constructed wetlands in Kenya and in the east African region are discussed.

References

Abira, M.A., van Bruggen, J.J.A., and Denny, P. 2005. Potential of a tropical subsurface constructed wetland to remove phenol from pre-treated pulp and papermill wastewater. *Water Science and Technology* 51 (9): 173 - 176.

Achoka, J.D. 1998. Levels of the physico-chemical parameters in the liquid effluents from Pan African Paper Mills at Webuye and in River Nzoia. PhD Thesis, Moi University, Kenya.

Akhtar, M. and Young, R.A. 1998. In: Young, R.A. and Akhtar, M. (eds.). *Environmentally friendly technologies for the pulp and paper industry.* John Wiley and Sons, inc., New York, pp. 1 - 2.

Azza, N.G.T., Kansiime, F., Nalubega, M. and Denny, P. 2000. Differential permeability of papyrus and *Miscanthidium* root mats in Nakivubo swamp, Uganda. *Aquatic Botany* 67: 169 - 178.

Batchelor, A. Scott, W.E., and Wood, A. 1990. Constructed wetland research programme in South Africa. In: Cooper, P.F. and Findlater, B.C. (eds.). *Constructed Wetlands in water pollution control.* Pergamon press, Oxford, UK, pp 373 - 382.

Batchelor, A. and Loots, P. 1997. A critical evaluation of a pilot scale subsurface flow wetland: 10 years after commissioning. *Water Science and Technology* 35 (5): 337 - 343.

Brix, H., 1993. Wastewater treatment in constructed wetlands: system design, removal processes, and treatment performance. In: Moshiri, G.A. (ed.). *Constructed wetlands for water quality improvement.* Lewis, Michigan, pp 9 - 22.

Byrd, J.F., Eysenbach, E.J. and Bishop, W.E, 1986. The effect of treated pulping effluent on a river and lake ecosystem. *Tappi Journal* (June) pp 94 - 98.

Chale, F.M.M. 1985. Effects of a *Cyperus papyrus L.* swamp on domestic wastewater. *Aquatic botany* 23: 185 - 189.

Chale, F.M.M. 1987. Plant biomass and nutrient levels of a tropical macrophyte (*Cyperus papyrus L.*) receiving domestic wastewater. *Hydrobiological bulletin* 21 (2): 167 - 170.

Cooper, P.F., Job, G.D., Green, M.B., and Shuttes, R.B.E. 1996. *Reed beds and constructed wetlands for wastewater treatment.* WRc plc. Swindon, Wiltshire, U.K., 184 pp.

Denny, P. 1985. Wetland vegetation and associated plant life forms. In: Denny, P. (ed.). *The ecology and management of African wetland vegetation.* Junk Publishers, Dordrecht, The Netherlands, pp 1 - 18.

Denny, P. 1997. Implementation of constructed wetlands in developing countries. *Water Science and Technology* 35 (5): 27 - 34

Edde, H., 1984. *Environmental control for pulp and paper mills.* Noyes Publications, Park ridge, New Jersey, USA. 500 pp.

Edde, H., 1994. Techniques for closing the water circuits in the pulp and paper industry. *Water Science and Technology* 29 (5-6): 11 - 18.

Elvers, B.; Hawkins, S., and Schulz, G. 1990. *Ullman's encyclopedia of industrial chemistry.* VCH, Germany.

Gapes, D.J., Frost, N.M., Clark, T.A., and Dare, P.H., 1999. Nitrogen fixation in the treatment of pulp and paper wastewaters. *Water Science and Technology* 40 (11-12): 85 - 92.

Gehm, H.W. and Bregman, J.I. (eds.) 1976. *Handbook of water resources and pollution control.* Van Nostrand Reinhold Company, New York, 840 pp.

GOK, 1999. *Kenya population census.* Ministry of Planning and National Development, Central Bureau of Statistics, Government of Kenya.

GOK/MENR, 1994. *Report of the Kenya National Environment Action plan.* Ministry of Environment and Natural Resources, Government of Kenya, Nairobi, 203 pp.

GOK/MWI, 2006. The National Water Resources Management Strategy 2006-2008. Ministry of Water and Irrigation, Government of Kenya.

Hammer, D.A. and Bastian R.K. 1989. Wetland ecosystems: natural water purifiers? In: Hammer D. A. (ed.), *Constructed wetlands for wastewater treatment: municipal, industrial and agricultural.* Lewis, Michigan, U.S.A., pp. 5 - 19.

Howard, G.W., 1992. Definition and overview. In: Crafter, S.A., Njuguna S.G., and Howard G.W. (eds.). *Wetlands of Kenya.* Proceedings of the Kenya Wetland Working Group Seminar on Wetlands of Kenya, National Museums of Kenya, Nairobi, IUCN, Gland, Switzerland, pp 1 - 4.

Howard, T.E., and Walden, C.C. 1974. Measuring stress in fish exposed to pulp mill effluents. *TAPPI* 57: 133 - 135.

JICA/GOK, 1992. *National Water Master Plan, Data Book: Hydrological data.* Japan International Cooperation Agency and Government of Kenya, Ministry of Water Development.

Johnston, P.A., Stringer, R.L., Santillo, D., Stephenson, A.D., Labounskaia, I. Ph. & McCartney, H.M.A. 1996. *Towards zero-effluent pulp and paper production: the pivotal role of totally chlorine free bleaching.* Greenpeace Research Laboratories Technical Report 07/96. Publ: Greenpeace International, Amsterdam, ISBN 90-73361-32-X. 33 pp.

Kansiime, F. and Nalubega, M. 1999. Wastewater treatment by a natural wetland: the Nakivubo swamp, Uganda: processes and implications. Ph.D. Thesis, A.A. Balkema Publishers, Rotterdam, The Netherlands, 300 pp.

Kipkemboi, J., Kansiime, F. and Denny, P. 2002. The response of *Cyperus papyrus* (L.) and *Miscanthidium violaceum* (K. Schum.) Robyns to eutrophication in natural wetlands of Lake Victoria, Uganda. *African Journal of Aquatic Sciences,* **27**: 11 - 20.

Kadlec, R.H., and Knight, R.L. 1996. *Treatment Wetlands.* Lewis Publishers, Boca Raton, Florida, 893 pp.

LVEMP, 1996. *Lake Victoria Environment Management Project document*

Mashauri, D.A., Mulungu, D.M.M. and Abdulhussin, B.S., 2000. Constructed wetland at the university of Dar es Salaam. *Water Science and Technology* 34 (4): 1135 - 1144.

NEMA, 1996. *State of the environment report for Uganda.* National Environment Management Agency (NEMA). http://easd.org.za/Soe/Uganda/CHAP4.html#4.4.1

Nemerow, N.L. 1978. *Industrial water pollution origins, characteristics and treatment.* Addison-Wesley Publishing Company, Reading, Massachusetts, 738p.

Okurut, T. O. 2000. A pilot study on municipal wastewater treatment using a constructed wetland in Uganda. PhD dissertation, Balkema publishers, Rotterdam, The Netherlands

PricewaterhouseCoopers, 2006. http://www.pwc.com/Extweb/industry.nsf/ accessed September 6, 2006.

Senzia, M.A.; Mashauri, D.A. and Mayo, A.W. 2002. Suitability of constructed wetlands and waste stabilisation pond in wastewater treatment: nitrogen transformation and removal. 3rd WaterNet/Warfsa symposium on water demand management for sustainable development, Dar es Salaam, October 2002.

Thompson, K., 1985. Emergent plants of permanent and seasonally flooded wetlands. In: Denny, P. (ed.) *The ecology and management of African wetland vegetation.* Junk Publishers, Dordrecht, The Netherlands, pp 43 - 107.

Thut, R. N. 1993. Feasibility of treating pulp and paper mill effluent with a constructed wetland. In: Moshiri, G.A. (ed), *Constructed wetlands for water quality improvement.* CRC Press, Florida pp 441 - 447.

UNEP, 1985. *Pollution abatement and control technology publication (PACT) for pulp and paper industry.* Industry and Environment Information Transfer Series, Paris.

Water Pollution Control Directorate, 1976. *Proceedings of Seminars on water pollution abatement technology in the pulp and paper industry.* Economic and technical review report EPS 3-WP-76-4 Environment Canada and Canadian Pulp and Paper Association, 220 pp.

Wetzel, R.G., 1993. Constructed wetlands: scientific foundations are critical. In: Moshiri, G.A. (ed.). *Constructed wetlands for water quality improvement.* Lewis, Michigan, pp 3 – 7.

Wood, A., and Hensman, L.C. 1989. Research to develop engineering guidelines for the implementation of constructed wetlands in Southern Africa. In: Hammer, D.A. (ed.), *Constructed wetlands for wastewater treatment: municipal, industrial and agricultural.* Lewis, Michigan, U.S.A., pp 581 - 589.

World Bank, 1996. World Bank/IFC project document on the modernisation of Pan African Paper Mills (E.A.) Ltd., Kenya.

Water Resources Management Rules, 2007. Kenya Gazette Supplement No. 92, Legislative Supplement No. 52, Legal Notice No. 171 of 28 September, 2007.

Chapter two

Quality of treated wastewater of Pan African Paper Mills (E. A.) Ltd

Quality of treated wastewater of Pan African Paper Mills (E.A.) Ltd.

Abstract

Pan African Paper Mills (E.A.) Limited (PANPAPER) is an integrated pulp and paper mill in Kenya that uses the Kraft process to make chemical pulp from wood and also recycles waste paper. The combined process wastewater is treated in aerated lagoons before discharge into the Nzoia River of the Lake Victoria basin. Limited historical data indicate that the final effluent seldom meets government-prescribed guidelines. This study was undertaken in order to identify and quantify pollutants in the wastewater that are problematic prior to the construction of a wetland for polishing the final treated effluent in order to buffer the river.

Treatment through the pond system resulted in effluent of a weaker strength than the raw wastewater. However, the final discharge into the Nzoia River, does not comply with the national discharge limits for BOD (30 mg/l), COD (100 mg/l), TSS (30 mg/l) and phenols (0.05 mg/l). The concentrations of pollutants in the final effluent averaged 45 ± 3, 394 ± 340, 52 ± 6, and 0.64 ± 0.09 for BOD, COD, TSS and phenols respectively. Total nitrogen and phosphorus concentrations (2.56 ± 0.40 mg/l and 0.74 ± 0.06 mg/l respectively) and bacterial populations (maximum 6×10^3 colony forming units per ml of sludge) were low while cyanophyta were the dominant algal type. The quality of the final effluent was highly variable and depended on the predominant processes in the factory. Pollutants concentrations were particularly high during maintenance dredging. There was a slight decrease in pH and dissolved oxygen, while the EC and total dissolved solids increased by 20 - 50% in River Nzoia downstream of discharge point when compared to an upstream location during low flow in the river. Organic matter (COD) increased by up to 100%, TSS by about 50% while phenols doubled in concentration. Nutrient ion concentrations also increased by 50 - 100%.

Significant improvements in internal control measures have been undertaken at PANPAPER Mills in the past three years reducing the level of lignin and chlorinated organic compounds in the process wastewater. Notwithstanding there is a need for alternative external measures to further purify the wastewater to ensure compliance with set regulations especially during low flow in the recipient Nzoia River in order to protect the ecosystem and public health. In a separate study we established the efficacy of a constructed wetland in purifying the wastewater.

Key words
Wastewater, PANPAPER Mills, pulp and paper mill, River Nzoia, aerated lagoons

Background

PANPAPER Mills commenced production of pulp and paper at the end of 1974 initially having a capacity of 45,000 metric tonnes of paper per year. This was gradually expanded and today the plant has a capacity of 150,000 tonnes per year. The average annual production at present is about 120,000 tonnes. The company supplies 80% of the domestic consumption of paper in Kenya. The factory uses the Kraft process to make chemical pulp and has a de-inking plant for wastepaper recycling. A detailed description of the processes is given in chapter one. Both bleached (70 tonnes per day) and unbleached (about 250 tonnes per day) grades of paper are made. Pulp is bleached using chlorine (60% as gas and 40% as sodium hypochlorite). The chlorine is produced at an electrolytic plant in the factory by the electrolysis of brine.

Fresh water for use at the factory is abstracted from the Nzoia River. Wastewater emanating from all processes in the factory is combined, screened and channelled through two parallel clarifiers each with a diameter of 48.8 m and a depth of 2.7 m. A commercial inoculum and fertilizer are dosed as the wastewater flows into the next treatment stage. The di-ammonium phosphate fertilizer is dosed at a rate of 150 kg/day. Up until October 2004, PANPAPER used aerobic bacterial seed. This has since been replaced by a seed composed of facultative microorganisms. The effluent is then treated through a series of lagoons (Figures 2.1 and 2.2) of which the first two are mechanically aerated. The design capacities of the ponds are presented in Table 2.1. The final treated effluent is discharged via a 200 m long channel into Nzoia River (Figure 2.3). At the upper end of the effluent channel is a steep cascade. This results in turbulent flow in the first half of the channel that is usually covered with foam. The effluent hydraulic residence time in the channel is quite short, approximately one hour. A luxuriant growth of vegetation, mainly *Typha domingensis* is found along the sides in the last 50 m section. The effluent is discharged into the Nzoia River via a diffuser outflow. The average daily discharge into the river is approximately 35,000 m^3 (factory records). Owing to the immensity of wastewater produced, discharge of inadequately treated mill effluent by PANPAPER is no doubt a priority environmental issue within the Lake Victoria basin.

Untreated pulp and paper mill effluents are known to contain toxic substances (Johnston *et al.*, 1996; Water Pollution Control Directorate, 1976) including chlorinated lignins and phenols, resins and resin acids; slimicides, and sulphur compounds. Such effluents can have deleterious effects in the environment to fish and other fauna (Howard and Walden, 1974; UNEP, 1985). Pulp and paper mill effluents are highly coloured with high loads of suspended solids that can reduce light penetration in the receiving water body thereby reducing primary productivity (Water Pollution Control Directorate, 1976; Byrd *et al.*, 1986).

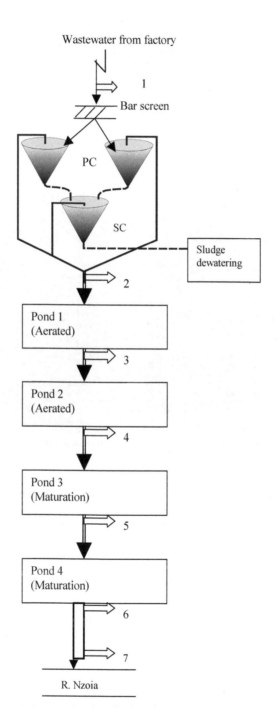

Fig. 2.1 Schematic diagram (not to scale) of PANPAPER mills' effluent treatment system. Block arrows indicate sampling points (1-7) PC = primary clarifier, SC = secondary clarifier

Fig 2.2 A view of PANPAPER Mills wastewater treatment ponds

Table 2.1 The design parameters and purpose of the PANPAPER ponds

	Water surface area (ha)	Water depth (m)	*Pond volume (m^3)	*Hydraulic Retention time (days)	Purpose
Pond 1 Aerated	7.37	3.0	221100	6.5	Aerobic break-down of organic matter
Pond 2 Aerated	4.33	3.0	129900	3.8	Aerobic break-down of organic matter
Pond 3 Stabilisation	11.86	3.5	415100	12.1	Settling of suspended solids -polishing
Pond 4 Stabilisation	3.72	3.5	130200	3.8	Settling of suspended solids - polishing

*The hydraulic retention times were calculated from pond volumes and the mean measured flow (34,201 m^3/day) at the outlet of Pond 4. It was assumed that the flow was uniform through all the ponds. The pond volumes were derived from dimensions obtained from the factory records.

Fig 2.3 The partially vegetated channel that conveys the final effluent into the Nzoia River

The Lake Victoria Environmental Management Project (LVEMP), which is responsible for the management of water quality in the basin, identified PANPAPER Mills as a site for an integrated tertiary industrial effluent treatment pilot project (LVEMP, 1996). This involved the construction of a pilot scale wetland to study the potential for polishing treated effluent in order to buffer the river. Studies have reported promising results with similar systems in temperate climates (Thut, 1993; Moore *et al.* 1994) but site-specific data are required to guide design and operation protocols (Kadlec and Knight, 1996). However, there are inadequate data on the nature and characteristics of the wastewater from PANPAPER Mills. Achoka (1998) examined heavy metal and alkali metal concentrations such as sodium and potassium in the final effluent. Data available at the factory and preliminary tests conducted in 1999-2000 by Abira *et al.* (2003) did not include regular measurements of dissolved oxygen (DO), biochemical oxygen demand (BOD), nutrients and toxic substances such as phenols. The objective of this study is to identify and quantify (characterise) pollutants in the wastewater of PANPAPER Mills prior to the construction of an artificial wetland for polishing the effluent before discharge to the Nzoia River. The main sampling survey of the ponds was from September to December 2002 (n = 6). Additional sampling was carried out between July 2003 and June 2005 (n = 4).

Materials and Methods

Physical and chemical quality characterization

In-situ measurements
In-situ measurements were recorded for pH, electrical conductivity (EC), DO, temperature and total dissolved solids (TDS) using a portable Multi-parameter Water Quality Monitor, model 6820-10M-0, manufactured by YSI Incorporated, USA. The measurements were taken at different reaches, including the overflows of the ponds just below the water surface. Effluent discharges were computed from flow measurements made using a current meter.

Organic matter, suspended solids and phenols
Wastewater was sampled every two to four weeks for the determination of BOD, chemical oxygen demand (COD), total suspended solids (TSS), Total nitrogen (TN) and total phosphorus (TP), phenols, sulphate and chloride at the following points: raw wastewater before screening (Fig. 2.1, sampling point 1), clarifier overflow (2), the overflows of Pond No.1 (aerated lagoon - 3), Pond No.2 (aerated lagoon - 4), Pond No.3 (stabilization lagoon - 5), and Pond No.4 (stabilization lagoon - 6). Samples were also taken at the end of the partially vegetated channel that conveys wastewater to the river. Laboratory analysis was carried out according to procedures described in the Standard Methods (APHA, 1995).

Nutrients, sulphates and chlorides
Total nitrogen and total phosphorus were determined on digested samples by cadmium reduction followed by nesslerisation (543 nm) and ascorbic acid (880 nm) methods respectively. Sulphate was analysed by the barium chloride turbidity method using a HACH turbidimeter model 26197-01 while chloride analysis was performed using a chloride meter Jenway model PCLM3.

Microbiological quality

Microbiological investigations were carried out to determine diversity of bacteria and algae in the ponds.

Bacteria were cultured in plate count agar according to *Standard Methods* (APHA, 1995). Smears on glass slides were examined under a compound microscope (Motic model 3906277 – B1 Series manufactured by System Microscopes). Identification of bacteria to species level was not possible.

Algal samples were taken twice from periphyton crusts floating on the surface of the ponds and the effluent channel. The crusts appeared to have been sloughed off the embankment by water movements. Dominant algal types were identified. Their relative abundance was scored using a three-point scale (abundant, frequent, and sparse). Identification followed the key described in *Standard Methods* (APHA, 1995). Both water and sludge samples from the stabilisation ponds and the effluent channel were taken for examination of fauna types.

Samples for chlorophyll *a* determination were taken in February and March 2002 from the stabilisation ponds and the effluent channel. From the ponds samples were collected from various locations namely: the inlet, opposite the inlet, at mid location (north and south banks) and near the outlet. At each site samples were collected in duplicate and filtered and preserved immediately for analysis later (*Standard Methods,* APHA, 1995).

Water quality in the Nzoia River

The river was sampled twice in February 2003 and six times in March/ April 2005 at locations upstream of the PANPAPER Mills discharge point (approx. 500 m) and downstream of discharge (about 3 km). Samples were analysed in the laboratory for BOD, COD, TSS, phenols, TN and TP as described for the factory wastewater. EC, pH, TDS and DO were measured in-situ.

Data analysis

Data generated were analysed using MS/Excel data analysis tools.

Results

General characteristics

General characteristics of the raw wastewater from PANPAPER Mills and the effluent leaving each stage of the treatment system are summarized in Tables 2.2 and 2.3. The raw wastewater quality is typical of a strong organic effluent. Treatment through the pond system resulted in effluent of a comparatively weaker strength.

In-situ measurements

The pH of the raw effluent was alkaline but after equalization in the clarifier it fell to below 7.5 in ponds 1 – 3, and then returned to a more alkaline value, 7.9 in pond 4 (Figure 2.4). Effluent dissolved oxygen remained low (less than 1 mg/l) throughout the treatment but rose to between 1 and 4.8 mg/l at station 7 just before discharge into the Nzoia River. A two-sample t-test revealed the rise in DO content was statistically significant ($p < 0.05$).). EC was high (2122 ± 209 µS/cm) and did not change significantly down the treatment stages (Figure 2.5). The temperature of the raw effluent from the factory is high (mean $41.5 \pm 1.0°C$) but cools to between 25°C and 34°C (ambient temperature) as it flows through the ponds.

Table 2.2 Wastewater quality ranges along the treatment system of PANPAPER Mills in Webuye – sampling points 1-7, n* = 10

Parameter/ Unit	Raw wastewater 1	Clarifier overflow 2	Pond 1 (Aerated) overflow 3	Pond 2 (Aerated) overflow 4	Pond 3 (Stabilisation) overflow 5	Pond 4 (Stabilisation) overflow 6	End of effluent Channel 7	Kenyan discharge guidelines‡
Temperature °C	38.3-44.6	38.4-42.6	33.3-35.1	27.3-30.6	25.4-31.2	23.3-29.1	26.5-28.1	
pH	6.1-9.1	7.7-8.7	7.0-7.6	7.1-7.7	7.4-7.5	7.7-8.2	7.9-8.0	6.5-8.5
EC mS/cm	1.39-2.66	1.43-2.86	1.68-2.38	1.66-2.28	1.46-2.11	1.56-2.30	1.52-2.26	
TDS g/l	0.70-1.27	0.72-1.79	0.84-1.33	0.84-1.34	0.73-1.38	0.78-1.39	0.76-1.38	
DO mg/l	1.2-3.0	0.20-0.52	0.0-0.61	0.0-0.35	0.0-0.22	0.06-0.36	0.94-4.87	
BOD5 (20°C) mg/l	92-633	127-424	109-292	83-153	44-63	32-52	41-60	30
COD mg/l	571-3085	440-963	429-1200	340-771	264-823	240-617	343-617	100
TSS mg/l	205-1265	54-177	100-210	73-138	44-94	24-78	20-66	30
Phenols mg/l	1.1-4.6	1.7-5.4	0.6-2.5	0.24-1.6	0.3-1.2	0.2-1.04	0.3-1.2	0.05
TN mg/l	0.19-3.5	0.7-3.0	1.7-3.3	1.6-3.3	0.3-3.5	0.41-4.1	0.4-2.4	10
TP mg/l	0.24-0.7	0.22-0.6	0.61-1.4	0.5-1.64	0.47-0.98	0.56-1.02	0.55-0.90	2
Sulphate mg/l	47-190	100-311	181-228	80-247	146-206	105-236	109-195	
Chloride mg/l	200-370	240-290	187-274	226-242	220-256	230-276	216-265	
Flow M³/day	-	-	-	-	-	32,158-36,245	-	

*Data collected from September to December 2002 (n = 6) Additional data collected between July 2003 and June 2005 (n = 4) were incorporated.   ‡= Water Resources Management Rules, 2007.

Table 2.3. Mean values for various parameters at each stage along PANPAPER Mills' effluent treatment system (n* = 10).

Parameter/ unit	Raw wastewater 1	Clarifier overflow 2	Pond 1 (Aerated) overflow 3	Pond 2 (Aerated) overflow 4	Pond 3 (Stabilisation) overflow 5	Pond 4 (Stabilisation) overflow 6	End of effluent Channel 7	Kenyan discharge guidelines
Temperature °C	41.5 ± 1.0	41.0 ± 1.3	33.9 ± 0.6	29.4 ± 1.1	27.6 ± 1.8	25.3 ± 1.3	27.3 ± 0.8	±3°C of ambient
pH	8.09 ± 0.59	7.84 ± 0.44	7.38 ± 0.17	7.30±0.22	7.46 ± 0.03	7.92 ± 0.12	7.95 ± 0.07	6.5-8.5
EC µS/cm	2122 ± 209	2318 ± 447	2123 ± 222	2019±185	1875 ± 208	1851 ± 199	1869 ± 170	
TDS mg/l	1286 ± 104	1390 ± 211	1203 ± 121	1166±113	1120± 198	1091 ± 152	1153 ± 139	
DO mg/l	2.51 ± 0.44	0.37 ± 0.09	0.22 ± 0.20	0.14±0.11	0.10 ± 0.06	0.22 ± 0.09	2.91 ± 1.97	
BOD5 (20°C) mg/l	249 ± 98	212 ± 54	152 ± 35	121 ± 10	54 ± 3	45 ± 3	49 ± 4	30
COD mg/l	1293 ± 1275	660 ± 579	608 ± 795	511 ± 498	465 ± 346	394 ± 340	445 ± 497	100
TSS mg/l	626 ± 126	92 ± 16	140 ± 17	113 ± 8	61 ± 6	52 ± 6	47 ± 9	30
Phenols mg/l	2.97 ± 0.39	2.92 ± 0.36	1.56 ± 0.27	0.83 ± 0.15	0.70 ± 0.08	0.64 ± 0.09	0.77 ± 0.16	0.05
TN mg/l	1.77 ± 0.29	2.02 ± 0.31	2.58 ± 0.23	2.42 ± 0.18	2.23 ± 0.44	2.56 ± 0.40	2.26 ± 0.66	10
TP mg/l	0.39 ± 0.06	0.31 ± 0.05	0.93 ± 0.10	0.96 ± 0.12	0.75 ± 0.05	0.76 ± 0.05	0.74 ± 0.06	2
Sulphate mg/l	135 ± 17	197 ± 32	206 ± 7	189 ± 18	179 ± 7	175 ± 15	158 ± 15	
Chloride mg/l	266 ± 19	260 ± 7	227 ± 16	236 ± 3	240 ± 8	242 ± 7	249 ± 9	
Flow m³/day	-	-	-	-	-	34,201±6988	-	-

*Data collected from September to December 2002 (n = 6) Additional data collected between July 2003 and June 2005 (n = 4) were incorporated. Effluent discharge guidelines for Kenya (Water Resources Management Rules, 2007) are included for comparison.

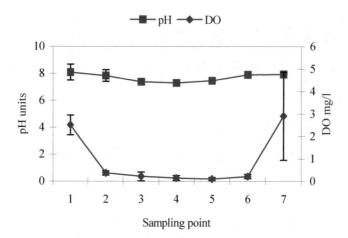

Fig. 2.4 Variation in pH and DO in the effluent treatment system at different sampling points (n=10). Note: Sampling points 1-7 respectively refer to: raw wastewater = 1, overflows from the clarifier = 2, pond 1 (aerated) = 3, pond 2 (aerated) = 4, ponds 3 and 4 (stabilisation) = 5 and 6, and the end of the effluent channel = 7. Data collected from September to December 2002 (n = 6). Additional data collected between July 2003 and June 2005 (n=4) were incorporated.

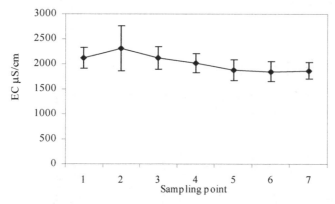

Fig. 2.5 Variation in EC in the effluent treatment system at different sampling points (n=10). Note: Sampling points 1-7 respectively refer to: raw wastewater = 1, overflows from the clarifier = 2, pond 1 (aerated) = 3, pond 2 (aerated) = 4, ponds 3 and 4 (stabilisation) = 5 and 6, and the end of the effluent channel = 7. Data collected from September to December 2002 (n = 6). Additional data collected between July 2003 and June 2005 (n=4) were incorporated.

Organic matter and suspended solids

In general the treatment system significantly modifies the quality of the raw wastewater (Figures 2.6 – 2.8). Overall, comparing the final effluent discharged (sampling point 6) with the raw wastewater (sampling point 1) concentration, reductions achieved were up to 92%, 82%, 69%, and 79% for TSS, BOD, COD and phenols, respectively.

TSS removal was negative at the outflows of the aerated ponds 1 & 2 when compared to the clarifier overflow. Most of the TSS and COD were removed at the clarifier. There was a strong correlation between COD and TSS removal throughout the system ($R^2 = 0.95$). The R^2- value was statistically significant (p-value = 0.00). BOD to COD ratios ranged from 0.12 to 0.32 with the lowest value being at the final effluent.

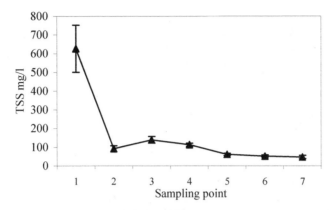

Fig. 2.6 Variation in TSS in the effluent treatment system at different sampling points (n=10). Note: Sampling points 1-7 respectively refer to: raw wastewater = 1, overflows from the clarifier = 2, pond 1 (aerated) = 3, pond 2 (aerated) = 4, ponds 3 and 4 (stabilisation) = 5 and 6, and the end of the effluent channel = 7. Data collected from September to December 2002 (n = 6). Additional data collected between July 2003 and June 2005 (n=4) were incorporated.

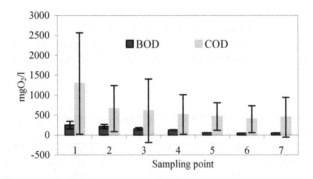

Fig. 2.7 Variation in BOD and COD in the effluent treatment system at different sampling points (n=10). Note: Sampling points 1-7 respectively refer to: raw wastewater = 1, overflows from the clarifier = 2, pond 1 (aerated) = 3, pond 2 (aerated) = 4, ponds 3 and 4 (stabilisation) = 5 and 6, and the end of the effluent channel = 7. Data collected from September to December 2002 (n = 6). Additional data collected between July 2003 and June 2005 (n=4) were incorporated.

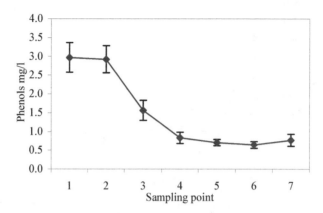

Fig. 2.8 Variation in phenols in the effluent treatment system at different sampling points (n=10). Note: Sampling points 1-7 respectively refer to: raw wastewater = 1, overflows from the clarifier = 2, pond 1 (aerated) = 3, pond 2 (aerated) = 4, ponds 3 and 4 (stabilisation) = 5 and 6, and the end of the effluent channel = 7. Data collected from September to December 2002 (n = 6). Additional data collected between July 2003 and June 2005 (n=4) were incorporated.

The mean percentage reduction in TSS, BOD, COD, and phenols at various stages of effluent treatment is given in Table 2.4. The end of the effluent channel (sampling point 7) has been included for completeness, as it was not designed for purpose of treating the effluent.

Table 2.4. Percentage reduction in TSS, COD, BOD, and phenols at various stages of wastewater treatment (Sampling points 2 – 7, in brackets)

Parameter mg/l	Clarifier overflow (2)	Pond 1 (Aerated) overflow (3)	Pond 2 (Aerated) overflow (4)	Pond 3 (stabilisation) overflow (5)	Pond 4 (stabilisation) overflow (6)	Effluent channel after macrophytes (7)
TSS	85	- 52	19	46	15	10
BOD	15	28	20	55	17	- 9
COD	49	8	16	9	15	- 13
Phenols	2	47	47	16	9	- 20

Note: percentage reduction refers to differences in inflow and outflow quality at each treatment stage. 'minus' indicates an increase in pollutant concentration.

The mean measured discharge of wastewater into Nzoia River was 34, 201 m³/day. This gives the loads of pollutants discharged per day into the river respectively as: 1.8 tonnes TSS, 13.5 tonnes COD, 1.5 tonnes BOD, and 0.02 tonnes phenols. Table 2.5 gives the mean discharged loads and derived concentration increase for different flow ranges in the Nzoia River. The increase is higher during low flow in the river.

Table 2.5 The resulting concentration increase from PANPAPER Mills effluent when discharged into the Nzoia River. The concentrations are in g/m^3 for all parameters except for phenols, TN and TP that are in mg/m^3

Parameter	Mean quantity discharged (g/s)	Concentration increase in Nzoia River		
		High (93 m^3/s)	Average (52 m^3/s)	Low (10 m^3/s)
TSS	21	0.22	0.4	2.1
BOD	17.8	0.19	0.34	1.8
COD	156	1.7	3	15.6
Phenols	0.25	2.6	4.8	25
TN	1	10	20	100
TP	0.29	3	5.4	29

Nutrients

The nitrogen and phosphorus content in the wastewater at each stage in the treatment system is presented in Figure 2.9. Both TN and TP concentrations rose after the clarifier (in Pond 1) where N and P fertilizer was added and declined in Ponds 2 and 3. Nitrogen levels rose slightly again at the outflow of Pond 4. The ratio of N to P was less than 3 in all the ponds except at the outflow of pond 4 where the ratio was 3.4. The BOD:N:P ratio in the final effluent was 100:5.7:1.7. The average nutrient loads discharged into the Nzoia River per day are 88 kg N and 26 kg P.

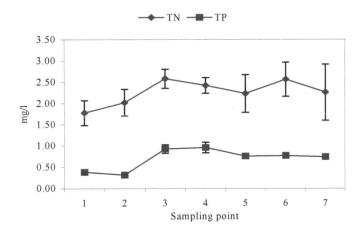

Fig. 2.9 Variation of TN and TP along the treatment system at different sampling points. Error bars for TP are omitted for clarity. Note: Sampling points 1-7 respectively refer to: raw wastewater = 1, overflows from the clarifier = 2, pond 1 (aerated) = 3, pond 2 (aerated) = 4, ponds 3 and 4 (stabilisation) = 5 and 6, and the end of the effluent channel = 7

Sulphates and chlorides

The concentrations of sulphates and chlorides were high throughout the treatment system (Table 2.2)

Microbiological characteristics

Total viable bacterial count
Bacterial counts of the water column were 3 - 5 times lower than in the sludge (Table 2.6). The counts in the sludge were four orders of magnitude less than that reported in the sludge of a primary clarifier of an activated sludge system in a Canadian bleached Kraft pulp and paper mill (Gauthier *et al.*, 2000). They were two to four orders of magnitude less than that found in domestic sewage activated sludge systems and tertiary lagoon effluents.

Table 2.6 Total viable bacterial counts (number of colony forming units per ml) at various sampling points (5 – 7).

Sample source	Sampling point	Water	Sludge	n
Raw wastewater	1	3600±3400	n.a	2
Clarifier overflow	2	50*	n.a	1
Pond 1 (aerated)	3	600	2000	1
Pond 2 (aerated)	4	950±550	6000±4000	2
Pond 3 (stabilisation)	5	643±87	2577±1378	3
Pond 4 (stabilisation)	6	617±159	3093±2015	3
Effluent channel	7	1180±478	2967±1454	3
Primary clarifier[1]		-	$2.2±0.2 \times 10^7$	
Secondary effluents[2]		-	1.1×10^6	16
Tertiary effluents[2]		6.6×10^4	-	11

Note: *Commercial inoculum dosed at outflow of clarifier, n.a = not applicable
[1](Gauthier *et al.*, 2000): primary clarifier of activated sludge system of a bleached Kraft pulp and paper mill. [2](Pike and Carrington, 1972): Domestic wastewater secondary effluents data are from nine filters and seven activated sludge plants while tertiary effluents are from ten lagoons and one grass plot

Types of bacteria
The colonies consisted mainly of long chains of spherical bacterial cells (cocci) and some rod-shaped ones. Samples from the vegetated channel downstream of the emergent aquatic macrophytes had, in addition, branched bacteria probably *Sphaerotilus*.

Algae
Cyanophytes were found to be predominant among the algal species observed in periphyton samples collected from the stabilisation ponds (Table 2.7, sampling points 5 and 6), and at the effluent channel after the macrophytes (sampling point 7). The periphyton was found on the stone embankment of the ponds at less than 5 cm below the water surface and on floating debris. Crusts sloughed off the embankment floating on the water surface included rotifers grazing on *Gomphosphaeria*, and ciliates.

Table 2.7. Algal types and relative abundance at the PANPAPER Mills stabilisation ponds and the effluent channel. Samples were collected in February and March 2002

Genus	Relative abundance	Phylum
Oscillatoria	++++	Cyanophyta
Anacystis	+++	Cyanophyta
Euglena	++	Flagellata
*Anabaena**	+	Cyanophyta
Gomphosphaeria	+	Cyanophyta
Agmenellum	+	Cyanophyta
Chlorogonium	+	Flagellata
Chlamydomonas	+	Flagellata
Phacus	+	Flagellata
Ankistrodesmus	+	Chlorophyta
Zygnema	+	Chlorophyta
Chlorella	+	Chlorophyta
Chlorococcum	+	Chlorophyta

Note: * present in samples taken in March 2002 only

Chlorophyll *a* concentrations in the stabilisation ponds and the effluent channel varied widely during the sampling period (Table 2.8). Pond surfaces were covered with effluent foam for most of March.

Table 2.8 Chlorophyll *a* concentrations (μg/l) for the PANPAPER Mills stabilisation ponds and the effluent channel (downstream of emergent aquatic macrophytes)

Sample source	Sampling date		
	February 2002	March 2002 (First week)	March 2002 (Last week)
Pond 3 (stabilisation)	95±3.6	50±1.8	14.6±1.8
Pond 4 (stabilisation)	104±4.8	52±2.0	15.4±1.2
Effluent channel	54±13	61±10	14.6±1.0

Water quality in the Nzoia River

Results obtained for water samples from the Nzoia River, both upstream and downstream of the effluent discharge point, are presented in Table 2.9 and Figure 2.10. There was a slight decrease in pH and dissolved oxygen, while the EC and dissolved solids increased by 20 % to 50 %. There was an increase in organic matter (COD) by up to 100 %, TSS by about 50 % while phenols generally doubled (100 %) in concentration. Nutrient ion concentrations also increased by 50 % to 100 %.

Table 2.9 Nzoia River water quality - mean values for *in-situ* measurements

	Sampling point	pH	DO mg/l	EC μS/cm	TDS mg/l
February 2003	Upstream	7.9	8.2	158	105
(n = 2)	Downstream	7.1	8	231	153
March 2005	Upstream	8.1	7.4	150	80
(n=5)	Downstream	7.9	6.8	170	110
April 2005	Upstream	7.9		150	80
(n= 3)	Downstream	7.6		200	120

Upstream = 500 m before PANPAPER Mill effluent discharge point; Downstream = 3 km after discharge point

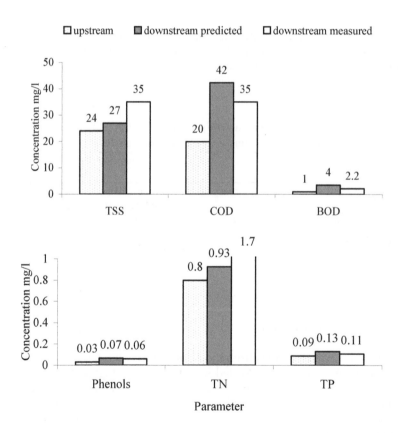

Fig. 2.10. Water quality of River Nzoia at sampling points upstream (500 m) and downstream (about 3 km) of PANPAPER's discharge point in March 2005. Predicted concentrations based on mean concentrations from the factory are included for comparison.

Discussion

The high electrical conductivity (EC) of PANPAPER Mills wastewater is due to the presence of chlorides and sulphates discharged into the effluent stream from the pulping and chlorination streams. Chlorides emanate from the plant that produces chlorine by electrolysis of brine. The wastewater is high in sodium, about 248 mg/l (Achoka, 1998) contributing to the high EC. The slight decrease in pH in ponds 2 – 4 may be due to conversion of sulphate in the ponds. Anaerobic conditions at the bottom of the ponds lead to some reduction of sulphates to hydrogen sulphide. The pH was within the range (6.5 – 8.5) suitable for carbonaceous oxidation (Gray, 2004).

The low level of oxygen in the aerated ponds is an indication that the ponds are only partially mixed, and hence facultative (Metcalf and Eddy, 2003). This allows some sludge to sediment on the bottom where it is partly decomposed by anaerobic conditions. The cascade at the outlet of pond 4 and the long sloping effluent channel aids aeration of the effluent and rise in the dissolved oxygen concentration just before final discharge.

Treatment system performance

The raw effluent exhibited high variation in quality due to differences in the main processes taking place at the mill. Effluent from the pulping process has higher organic matter and residuals from the digester (black) liquor while the effluent from the paper-making process contains mainly inorganic substances (white water) added as mineral fillers, dyes, defoamers, and slimicides (Nemerow, 1978).

The clarifiers' performance was variable 75 - 96 % (average 85 %) and sometimes below the designed objective of 90 % sediment and TSS removal. This variation is caused in part by the presence of floatable material from the recycled paper plant. Small shreds of plastics that passed through the bar screen were seen floating on the clarifier water surface. Paper recycling plants are known to release high TSS from fillers, coatings, tines, inks, glue and plastic (Wallendahl, 1995).

The aerated ponds were designed to degrade organic matter biologically. Lagoons are capable of removing degradable organic matter (BOD) to the tune of 80% for aerobic lagoons and up to 95 % for facultative ponds (Eckenfelder, 1980). However, the generated biomass has to be settled externally, in a post-settlement pond (Metcalfe and Eddy, 2003; Eckenfelder, 1980). In the PANPAPER lagoons floating aerators tended to keep the in-situ generated biomass partly in suspension. This together with the ponds' de-sludging probably caused the increase in TSS in ponds 1 and 2. The organic solids were mainly removed by settling in the first stabilization pond thereby removing the associated BOD (55 %). Overall, the aerated lagoon effluent had BOD in the range 83-153 mg/l. This is comparable with the normal range of 80 - 200 mg/l BOD for similar ponds (Eckenfelder and Ford, 1970).

COD removal was highly correlated with TSS removal indicating that sedimentation was the main mechanism responsible for removal. The achieved removal rate of 69 % is comparable to that reported for a similar system in New Zealand (62 %, Stuthridge et al., 1991). Pulp and paper mill wastewater consists of cellulose fibres that exert an oxygen demand measurable as COD but are not readily degradable so they do not contribute to the immediate oxygen demand measurable by BOD tests (Edde, 1984). The COD also consists of dissolved, hard to degrade, lignins and lignin degradation products. The low BOD: COD ratio (about 1:10) in the final effluent is an indication of such substances.

Phenol removal occurred mainly in the aerated ponds (biological stage). A very small proportion was removed at the clarifier indicating that adsorption is a minor removal mechanism. Eckenfelder (2000) reports that phenol removal is mainly microbial-mediated (up to 70 %) with adsorption and other processes accounting for the remainder. The overall removal achieved (79 %) is similar to that reported for chlorophenols in general in the New

Zealand system mentioned above (Stuthridge *et al.*, 1991) with higher removal rates (up to 84 %) for specific chlorophenols.

Pulp and paper mill wastewaters are generally deficient in nutrients and supplementation is normal to facilitate biological treatment after clarification (Edde, 1984; Gapes *et al.*, 1999). The addition of fertilizer at the clarifier outflow at PANPAPER Mills is reflected in nitrogen and phosphorus increases in the effluent stream. The amount of nitrogen is probably insufficient, as the ratio of N: P in the wastewater leaving pond 1 was only 2.5 and further consumption by biological activity in the second aerated lagoon reduced the ratio further (N: P ratio 2:1). The slight increase in the final stabilization pond (N: P ratio 3.4:1) was probably due to nitrogen fixation by microorganisms and benthic recycling (Slade *et al.*, 1999; Gapes *et al.*, 1999). Nitrogen fixation by bacteria has also been reported in primary clarifiers of activated sludge systems treating pulp and paper mill wastewater (Gauthier *et al.*, 2000). The predominance of cyanophytes in our stabilisation ponds suggests nitrogen fixation. A less P-containing fertilizer mixture may increase the N:P ratio. Adjusting the dosage of N may also improve the amount of N for BOD degradation (see below).

The low counts of bacterial populations in the PANPAPER Mills aeration and stabilisation ponds can be attributed to a variety of reasons. First, the level of nutrients in the ponds is generally low. In general, the nitrogen requirement for microbial metabolism is satisfied if the BOD:N ratio is less or equal to 18:1 (Gray, 2004). Gray (2004) asserts that at higher ratios the metabolism is much less efficient. The ratio was 24:1 and 18:1 respectively for the stabilisation ponds 3 and 4 respectively. This may have created an environment for slow growth resulting in low bacterial numbers. The second reason is the effluent toxicity and variation in the wastewater quality that could affect the acclimation of the microorganisms (Werker and Hall, 1999). Finally, if conditions such as pH are not sufficient for their degradation, biocides used for slime control could cause intermittent acute toxicity of final effluents (Milanova and Sithole, 1997). Similar conditions as well as turbidity and effluent foam may have caused the declining chlorophyll *a* concentration. The chlorophyll *a* concentrations measured in late March 2002 were low (15 µg/l), indicating low algal productivity in the ponds compared to that of a more productive system such as the Lake Victoria. The Lake has been reported to have concentrations in excess of 40 µg/l (Kling *et al.*, 2001) in recent years. The presence of foam in PANPAPER Mills ponds, besides obstructing light, may concentrate toxic substances especially resin acids (Water Pollution Control Directorate, 1976).

Implication of PANPAPER Mills discharges for the water quality of the Nzoia River

The concentrations of potential pollutants in the final effluent (Table 2.3) exceed that allowable by the regulating authority in Kenya. Each cubic metre of river water that flows past the discharge point every second acquires an increase in concentration depending on the flow in the river (Table 2.5). The increase is highest during low flow in the River, usually in the dry season (January to March). This was reflected in an increase in the concentration of various parameters by between 20 and 120 % at the downstream sampling point during the month of March (Figure 2.10).

The effects of the pollutants are manifold. Suspended solids mainly composed of cellulose fibres blanket the riverbed and exert an oxygen demand up to double that which may be exerted by municipal sludge whilst soluble organics cause depletion of oxygen in the receiving stream (Eckenfelder, 1980; 2000). Phenols are toxic to fish and combination with chlorine they impart a bad taste to drinking water at low concentrations of 0.005 mg/l (Dickson *et al.*, 1978). This implies that additional water treatment, at extra cost, is required to remove phenols and other organics.

The concentrations of TN and TP discharged by PANPAPER Mills are low (less than 3 mg/l and 1 mg/l respectively). They are less than the new Kenyan discharge guidelines

(Water Resources Management Rules, 2007) as well as the European Union Urban Wastewater Treatment Directive (91/271/EEC - TP < 2 mg/l, TN < 10 mg/l). The latter is usually considered stringent (Gray, 2004) for discharge into surface waters classified as sensitive or at risk of eutrophication. The concentrations of TN in the Nzoia River at low flow are 50 –100 % more than pre-discharge levels. Due to the large quantity of effluent discharged, there is a potential for enrichment especially in impoundments downstream, including the Lake Victoria.

Conclusion

The present effluent treatment system at PANPAPER Mills is performing well as per its type and design. However, the final effluent discharged into the Nzoia River, despite being much weaker than the raw wastewater does not comply with the national discharge limits for BOD (30 mg/l), COD (100 mg/l), TSS (30 mg/l), and phenols (0.05 mg/l).

There is therefore need to undertake additional control measures both internal and external. Significant improvements in internal control measures were undertaken from 2002 to 2004 (Personal communication with the Technical Manager of PANPAPER Mills). For example, the modification of the batch digesters to allow for extended delignification to a lower kappa number and part substitution (40 %) of molecular chlorine used in bleaching. This reduced the level of residual lignin and chlorinated organic compounds. Despite such controls alternative external measures are necessary to ensure compliance with set regulations especially during low flow in the receiving Nzoia River to protect the ecosystem and public health. The potential of an alternative cost-effective technology such as a constructed wetland in buffering the river will be studied.

References

Abira M. A, Ngirigacha H. W. and van Bruggen J.J.A. 2003. Preliminary investigation of the potential of four tropical emergent macrophytes for treatment of pre-treated pulp and papermill wastewater in Kenya. *Water Science and Technology* 48 (5): 223 - 231.

Achoka, J.D., 1998. Levels of the physico-chemical parameters in the liquid effluents from Pan African Paper Mills at Webuye and in River Nzoia. PhD Thesis, Moi University, Kenya.

APHA, 1995. *Standard methods for the analysis of water and wastewater*, 19th edition. American Public Health Association, Washington DC.

Byrd, J.F., Eysenbach, E.J. and Bishop, W.E. 1986. The effect of treated pulping effluent on a river and lake ecosystem. *Tappi Journal* (June) pp. 94 - 98.

Dickson, K.L., Cairns Jnr, J. and Livingston, R.J. (eds.) 1978. *Biological data in water pollution assessment: Quantitative and statistical analyses* STP6522. American Society for Testing and Materials.

Eckenfelder, W.W. and Ford, D.L. 1970. *Water Pollution Control-Experimental procedures for process design*. Pemberton press.

Eckenfelder, W.W. 1980. *Principles of water quality management*, CBI publishing. Boston, Massachusetts, 717pp.

Eckenfelder, W. W. Jr. (2000). *Industrial Water Pollution Control*, 3rd edition, McGraw-Hill, New York U.S.A.

Edde, H. 1984. *Environmental control for pulp and paper mills*. Noyes Publications, Park ridge, New Jersey, USA. 500p.

Gapes, D.J., Frost, N.M., Clark, T.A., and Dare, P.H. 1999. Nitrogen fixation in the treatment of pulp and paper wastewaters, *Water Science and Technology* 40 (11-12): 85 - 92.

Gauthier, F., Neufeld, J.D., Driscoll, B.T. and Archibald, F.S. 2000. Coliform bacteria and nitrogen fixation in pulp and paper mill effluent treatment systems. *Applied and Environmental Microbiology* 66 (12): 5155 – 5160.

GOK, 1999. *Kenya population census*, Ministry of Planning and National Development, Central Bureau of Statistics, Government of Kenya.

Gray, N.F. 2004. *Biology of wastewater treatment*, second edition, Imperial College press 1421 pp. London. U.K.

Howard, T.E., and Walden, C.C. 1974. Measuring stress in fish exposed to pulp mill effluents. *TAPPI* 57: 133 - 135.

JICA/GOK, 1992. *National Water Master Plan, Data Book: Hydrological data*. Japan International Cooperation Agency and Government of Kenya, Ministry of Water Development.

Johnston, P.A., Stringer, R.L., Santillo, D., Stephenson, A.D., Labounskaia, I. Ph. & McCartney, H.M.A. 1996. *Towards zero-effluent pulp and paper production: the pivotal role of totally chlorine free bleaching*. Greenpeace Research Laboratories Technical Report 07/96. Publ: Greenpeace International, Amsterdam, ISBN 90-73361-32-X. 33pp.

Kadlec, R.H., and Knight, R.L. 1996. *Treatment Wetlands*, Lewis Publishers, Boca Raton, Florida, 893 pp.

Kling, H.J., Mugidde, R. and Hecky, R.E. 2001. Recent changes in the phytoplankton community of Lake Victoria in response to eutrophication. *The Great Lakes of the World (GLOW): Food-web, health and integrity:* 47 - 65, Backhuys Publishers, Leiden, The Netherlands.

LVEMP, 1996. *Lake Victoria Environment Management Project document*

Milanova, E. and Sithole, B.B., 1997. Acute toxicity to fish and solution stability of some biocides used in the pulp and paper industry. *Water Science and Technology* 35 (2-3): 373 - 380.

Metcalf and Eddy, Inc. 2003. *Wastewater engineering: treatment disposal and re-use*, 4th edition. Revised by G. Tchobanoglous. F.L. Burton and H.D. Stensel, McGraw-Hill, New York, USA, 1819p.

Moore, J.A., Skarda, S.M. and Sherwood, R. 1994. Wetland Treatment of pulp and paper mill wastewater. *Water Science and Technology* 29(4): 241 - 247.

Nemerow, N.L. 1978. *Industrial water pollution origins, characteristics and treatment*. Addison-Wesley Publishing Company, Reading, Massachusetts, 738p.

Pike, E.B. and Carrington, E. G. 1972. Recent developments in the study of bacteria in the activated-sludge process. *Water Pollution Control* 71: 583 - 605

Slade, A.H., Nicol, C.M. and Grigsby, J. 1999. Nutrients within integrated bleached Kraft mills: sources and behaviour in aerated stabilization basins. *Water Science and Technology* 40 (11-12): 77 - 84.

Stuthridge, T.R., Campin, D.N., Langdon, A.G., Mackie, K.L., McFarlane, P.N. and Wilkins, A.L. 1991. Treatability of bleached Kraft pulp and paper mill wastewaters in a New Zealand aerated lagoon treatment system. *Water Science and Technology* 24 (3/4): 309 - 317.

Thut, R. N. 1993. Feasibility of treating pulp and paper mill effluent with a constructed wetland. In: Moshiri, G.A. (ed), *Constructed wetlands for water quality improvement*, CRC Press, Florida pp 441 - 447.

UNEP, 1985. *Pollution abatement and control technology publication (PACT) for pulp and paper industry*. Industry and Environment Information Transfer Series, Paris.

Wallendahl, U, 1995. Pulp and Paper industry. In: Higgins, T.E. (ed). *Pollution prevention handbook* Lewis Publishers, Boca Raton, pp 513 - 528.

Water Pollution Control Directorate, 1976. *Proceedings of Seminars on water pollution abatement technology in the pulp and paper industry*. Economic and technical review report EPS 3-WP-76-4 Environment Canada and Canadian Pulp and Paper Association, 220pp.

Werker, A.G. and Hall, R.E. 1999. Limitations for biological removal of resin acids from pulp mill effluent. *Water Science and* Technology 40 (11-12): 281 - 288.

Water Resources Management Rules, 2007. Kenya Gazette Supplement No. 92, Legislative Supplement No. 52, Legal Notice No. 171 of 28 September, 2007.

Chapter three

The design, hydrology and flow characteristics of the pilot-scale constructed wetland

The design, hydrology and flow characteristics of the pilot-scale constructed wetland

Abstract

The pilot-scale constructed wetland at Pan African Paper Mills (E.A.) Ltd. (PANPAPER) in western Kenya consisted of eight subsurface flow cells each of dimensions 3.2 m (length) × 1.2 m (width) × 0.8 m (depth) and two cells of dimensions 6.2 m (length) × 1.5 m (width) × 0.8 m (depth). The latter were initially operated as free water surface flow and later as subsurface flow systems. The subsurface flow cells were planted in pairs with *Cyperus immensus*, *Typha domingensis*, *Phragmites mauritianus* and *Cyperus papyrus* respectively. The *Cyperus immensus* were removed after eight months and the cells left unplanted. The larger cells were planted with *Typha domingensis*. All cells were filled with gravel to a depth of 30 cm and had an impermeable barrier that excluded seepage and infiltration. The systems operation was dynamic involving different operation modes, hydraulic loading rates and retention times in order to optimise pollutant removal while maintaining good plant vitality. In all there were three phases of batch operation and two of continuous flow.

Rainfall, evapotranspiration (ET), flow rates and tracer (lithium chloride) flow patterns were determined to provide a basis for the evaluation of the wetland systems' performance and water budget. Evapotranspiration (8 -16 mm/day) was found to be an important component of outputs in the water budget of the wetland system making up 15 – 32 % depending on system type. It should therefore be an integral part in wetland design in the tropics. ET rates were different for different aquatic plant species. Cells planted with *Typha* domingensis had the highest rates. For similar plant species, the rates seemed to depend on shoot density and plant biomass. ET rates were higher in continuous flow compared to batch loading with daily compensation.

The actual retention time and other hydraulic parameters (efficiency and number of "tanks in series") could not be determined as there was no discernable tracer concentration curve for all wetland cells. However, it may be deduced from the sharp spikes that some measure of plug flow did take place albeit following several micro channels. For pulp and paper mill wastewater, which has high organic matter content, the study should be conducted with a different tracer. Alternatively, lithium chloride may still be used but with continuous feed instead of pulse feed, as was the case in this study. Determination of the adsorption capacity of accumulated sludge is necessary in determining the tracer dosage into the wetland.

Key words

Wetland, design, evapotranspiration, flow characteristics, water budget, hydrology

Introduction

The Pan African Paper Mills (E. A.) Ltd. (PANPAPER) in Webuye, western Kenya is an integrated pulp and paper factory that uses mainly wood to make paper and board. The factory produces large quantities of wastewater (average 35,000 m^3/day). The wastewater is channelled through a bar screen into two parallel clarifiers then treated through a series of ponds two of which are aerated (Chapter 2). The pond system although operating optimally yields a final effluent that does not meet national discharge guidelines. Besides additional internal control measures undertaken in the period 2002 - 2004 there is still need for alternative cost-effective external control measures to improve the quality of the final effluent and buffer the recipient River Nzoia. Under the Lake Victoria Environment Management project it was decided to explore the use of a constructed wetland as a tertiary stage in polishing the effluent. This chapter presents the structure of a pilot-scale study constructed wetland and discusses the role of evapotranspiration in the water balance of such a system under tropical climatic conditions. An attempt was made to determine the wetland's flow characteristics.

Wetland system design

Constructed wetland systems can be considered as attached growth biological reactors, and their performance can be described with first order plug-flow kinetics if steady state conditions are assumed (US EPA, 1988; Watson et al., 1989). According to Metcalf and Eddy (1991), organic matter (BOD, COD, total organic carbon) degradation, nitrification, disinfection, and adsorption generally follow first order kinetics. However, Kadlec et al. (1993) caution that due to variability in wetland characteristics, for instance, depth and hydraulic loading that depend on hydrologic factors, presumption of plug flow kinetics is erroneous and that the design equations based upon it are hence incorrect. Moreover, lack of information on the operative mechanisms on lump-sum parameters such as BOD, and suspended solids limits the use of design equations based on them (Tchobanoglous, 1993). This is because the nature of the residual organic matter that causes BOD is continuously changing. For suspended solids there is no information available on particle size distribution.

Assumption of plug flow conditions is used only to provide an approximation of prevailing conditions in the wetland. It has been used to fit data obtained from organic matter removal (BOD) in batch-fed wetlands (Kadlec and Knight, 1996). Other design models employed depend on multiple regression analysis of performance data from operating systems (Kadlec, 2000). Others assume that the biological reactions that occur in the wetlands are similar to those of attached growth wastewater treatment processes. All approaches can lead to valid results if applied properly (Reed et al., 1995).

In this study the wetland design followed the criteria given by Kadlec and Knight (1996) and WPCF (1990) for both free water surface flow (FWS) and horizontal subsurface flow (HSSF) constructed wetlands. Previous work on wetland systems treating pulp mill wastewater included both subsurface flow (e.g. Thut, 1993; Moore et al., 1994) and surface flow systems (e.g. Hammer et al., 1993, Boyd et al., 1993). Subsurface flow systems have proved to be more effective, giving higher efficiency pollutant removal per unit area, and are compact (Cooper et al., 1996). In the warm tropics they do not give problems with mosquito breeding. The main system of choice in this study was therefore a gravel bed subsurface flow. However, a free water surface flow (FWS) system was included in a parallel study for purposes of comparison. Moreover, it was thought that the high variability in wastewater quality (Chapter 2) may necessitate the use of a hybrid system, using the FWS system to reduce the suspended and organic matter loading into the HSSF system as suggested by Kadlec (2005).

Wetland hydrology (water balance)

Wetland systems interact strongly with the atmosphere via rainfall and evapotranspiration (Kadlec, 1989). Evapotranspiration (ET) occurs with strong diurnal and seasonal cycles (Kadlec and Knight, 1996). Both parameters combine to either dilute or concentrate wastewater in the wetland. High vegetation density influences water movement along the wetland and increases evapotranspiration. By lengthening or shortening detention time, the water mass balance influences pollutant mass removal. For a constructed wetland the water balance can be expressed as follows (US EPA, 1988):

$$Q_i - Q_o + P - ET = \frac{dV}{dt} \qquad \text{Equation 3.1}$$

Where:

Q_i	=	influent wastewater flow, volume/time
Q_o	=	effluent wastewater flow, volume/time
P	=	precipitation, volume/time
ET	=	evapotranspiration, volume/time
V	=	volume of water
t	=	time
dV/dt	=	change in wetland storage

Groundwater inflow and infiltration are excluded from the equation since the constructed wetland studied had an impermeable barrier.

Flow characteristics

Wetlands are frequently characterized by non-ideal flow patterns. The existence of micro channels in the wetland bed determines the flow of water and transport of pollutants hence the actual detention time (Kadlec and Knight, 1996). The actual detention time can be determined by use of a non-reactive tracer. In this study lithium chloride was used to determine the actual retention time during the continuous flow mode of operation.

The tracer concentration-time distribution can give additional information on the nature of flow or extent of mixing within the wetland system. Kadlec and Knight (1996) analysed the tracer response for FWS wetlands and concluded that a "tanks-in-series" (TIS) model with three "continuously stirred tank reactors" (CSTRs) would best represent the observed flow characteristics. The degree of mixing determines the number of CSTRs. The number of CSTRs in series (N) may be calculated from the ratio of the nominal detention time (t_n) to the difference between the nominal detention time and the time of peak outflow concentration (t_p) i.e. $N = t_n/(t_n - t_p)$.

A tracer peak with a very small standard deviation about the mean residence time is characteristic of plug flow conditions. Subsurface flow wetlands, however, present a time delay (t_d) before tracer outflow is observed and the flow hydrodynamics may be described by a combination of plug flow and CSTRs. The presence of micro-channels, shortcuts and re-circulation zones lead to a loss in effective detention volume thus reducing the hydraulic efficiency (Persson et al., 1999). They developed a measure for the hydraulic efficiency (λ) that reflects tracer flow characteristics of constructed wetlands and ponds. In its simplest form, λ is the ratio of the time of the peak outflow concentration (t_p) to the nominal detention time (t_n).

Research scope and objectives

The pilot study constructed wetland was intended to act as a polishing unit for the wastewater from the PANPAPER Mills stabilisation ponds and to offer continuous buffer, especially during low flow in the River Nzoia. The objectives of this study therefore were: (1) To design a wetland system that would improve the PANPAPER Mill wastewater to meet set national guidelines. (2) To determine the water balance and flow characteristics of the constructed wetland system.

Materials and Methods

The approach used was:
1) To design a small-scale pilot wetland using established criteria for HSSF and FWS systems. The construction was done in such a manner as to allow for replication of units (cells), sampling ports, and flexibility in loading regime in order to facilitate the determination of optimal operating conditions for removal of organic matter, suspended solids, nutrients and phenols.
2) To use four locally occurring aquatic plants, namely *Cyperus immensus, Cyperus papyrus, Typha domingensis* and *Phragmites mauritianus*. These were planted in gravel of trachyte origin. The gravel would improve hydraulic conductivity, minimize clogging and permit negligible adsorption of phosphorus.
3) To measure water loss due to evapotranspiration.
4) To determine flow characteristics using a tracer (lithium chloride) under the continuous flow mode of operation.

Wetland design, construction and start up

The appropriate wetland sizes were derived based on criteria given by WPCF (1990) and as described by Kadlec and Knight (1996), the available pilot study land area, and the wastewater quality characteristics (influent BOD_5 maximum 100 mg/l and effluent BOD_5 of 20 mg/l). Four pairs of rectangular-shaped cells covering a total surface area of 30.7 m^2 were constructed for the subsurface flow system. The cells were operated in parallel. Each was 1.2 m (width) x 3.2 m (length) x 0.8 m (depth) in size. The pairs were initially planted with *Cyperus immensus* (series A), *Typha domingensis* (series B), *Phragmites mauritianus* (series C) and *Cyperus papyrus* (series D) respectively. The general layout of the cells is presented in Figure 3.1. One pair of larger cells of a FWS system with dimensions 1.5 (width) x 6.2 m (length) x 0.8m (depth) planted with *Typha domingensis* (series E) was included for comparison.

Construction commenced in December 2001 and was completed at the end of June 2002. The cells were constructed with concrete blocks. The bed was lined with heavy-duty polyethylene sheets to prevent seepage or infiltration, and had a slope of 1 %. All the cells were filled to a depth of 30 cm with washed gravel of trachyte origin with uniform size 6.25 mm (at least 95 %). The gravel was deeper at the outlet end by 1% of inlet depth. Both ends were filled with larger size gravel (ca 40 mm). A 90° V-notch weir was installed for measuring influent flow into each cell. A perforated unplasticised polyvinyl chloride (uPVC) pipe was fitted along the width of each cell of the SSF system at height of 5 cm and another at 15 cm from the floor bed for sampling. The perforations were made at 50 mm centres and had a diameter of 10 mm. The effluent was collected via a 75 mm perforated pipe. A flexible hosepipe was fitted to the outlet pipe in the drain chamber. The hose was used to regulate the water level in each cell. The water level was maintained at 28 cm. A 50 mm-drainage pipe for washout was fitted on the floor of each cell.

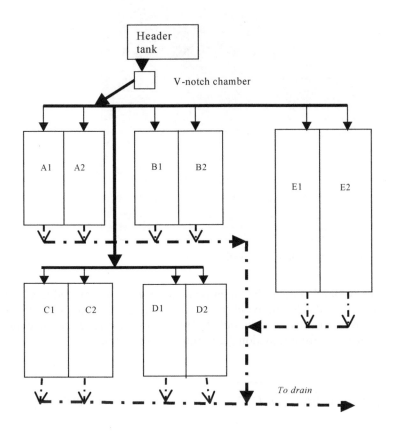

Fig 3.1 General layout plan of constructed wetland (not to scale). Wetland cell pairs A - D are subsurface flow of equal size (3.7 m^2) while pair E is free water surface flow. A1, A2 are control (unplanted); B1, B2 and E1, E2 = *Typha domingensis*, C1, C2 = *Phragmites mauritianus*, D1, D2 = *Cyperus papyrus*. The A-series had *Cyperus immensus* in the first 8 months.

The wetland cells were filled with river water prior to planting. Young plants of *Cyperus immensus* and *Phragmites mauritianus* were transplanted into the wetland cells from a nursery at 30 cm centres (density of 7.2 shoots/m^2). Healthy young plants of *Cyperus papyrus* and *Typha domingensis* collected from natural wetlands were planted directly into the wetland cells. The plants were left in river water for one week. The river water was partially mixed with wastewater from the final stabilisation pond of PANPAPER Mills over the following two weeks to enable plants to acclimatise. Thereafter plants were left to establish in wastewater for three months (prior to actual experimental treatments) when most had attained a height of 1.0 – 1.5 m except *C. immensus*, which was about 0.3 - 0.5 m in height. Figure 3.2 shows the front view of the established wetland system

Fig. 3.2 Front view of the established wetland system

In order to calibrate the wetland system, the outlet pipes were set at the desired level. The flow in the V-notch chamber was adjusted to avoid turbulence while a stopwatch recorded the time taken for the first few drops to emerge at the outlet. The process was repeated once more. The flow in the cells was calculated from Equation 3.2 (Young *et al.*, 1997).

$$Q = C_{wt} \times \frac{8}{15} \tan\left(\theta/2\right) \sqrt{2gH}^{5/2}$$ Equation 3.2

Where:
Q   = flow (m³/s), H = vertical height of water (m)
g   = gravitational acceleration (9.81 m/s²)
θ   = V- notch angle (90°)
C_{wt} = the weir coefficient for triangular shaped V-notch weir
       (for H < 0.06m, C_{wt} = 0.6)

The resulting interstitial volume was obtained by multiplying the flow by the time (seconds) taken to fill each cell to the preset level. By mid 2004, after nearly two years' operation, the wastewater had corroded the V-notch weirs thereby altering the notch angle. The wetland cells were henceforth calibrated periodically by completely draining then refilling with known amounts of wastewater. The wastewater-loading regime was manipulated in an attempt to achieve the goal of optimal treatment efficacy.

Initially the wetland was operated on a batch flow regime. Preliminary experiments in bucket mesocosms (Abira *et al.*, 2003) revealed that wetland plants were healthier and greener in treatments at higher hydraulic loading rates (HLR) with a hydraulic retention time (HRT) of five days, than in treatments at lower HLR and retention time of 10 days. In this study, therefore, the wetland cells were initially operated at 5 days retention time followed by 3 days retention. The wetland was recalibrated periodically to determine the new total volumes as plant root and rhizome biomass increased (Table 3.1).

Wetland system operation

Batch loading Phase 1

5-day HRT - November 2002 to July 2003. The loading of wetland cells A1 & A2 commenced only on 24th December 2002 due to poor establishment of plants. Influent wastewater was loaded and drained batchwise every five days in all ponds. Water loss due to evapotranspiration was compensated by topping up with pre-settled river water between 8.00 am and 9.00 am daily. Wastewater was sampled at the inlet and outlet of the cells from about 11.00 hrs.

Batch loading Phase 2

3-day HRT – July 2003 to March 2004. This was conducted in a similar manner to phase1. However, due to patchy establishment of plants in series A cells it was decided to remove all plants as well as any dead roots from the cells and to leave them as unplanted controls.

Batch loading Phase 3

5-day HRT - February to April 2005. During the first phase of the wetland batch operation it was found that the subsurface wetlands performance (series A – D) was less efficient compared to that at 3-day HRT (Abira *et al.*, 2005). One of the reasons may be that initially the wetland systems' microbial environment was not yet fully established. The purpose of repeating this mode of operation therefore was to determine if the performance would be different with the mature wetland system. The system operation in this mode lasted 75 days and involved the unplanted cells, *Phragmites* and *C. papyrus* cells.

Table 3.1 Wetland volumes in batch mode operation. The instantaneous hydraulic loading rate was maintained between 2 - 4 cm/day. The cells were drained slowly for 2 - 3 hours

Wetland Cell No	Cell type	Volume per batch, m^3		
		Phase 1 5 days HRT	Phase 2 3 days HRT	Phase 3 5 days HRT
A1	Unplanted*	0.417	0.426	0.410
A2	Unplanted*	0.401	0.433	0.426
B1	*Typha*	0.479	0.404	**
B2	*Typha*	0.472	0.397	**
C1	*Phragmites*	0.485	0.435	0.417
C2	*Phragmites*	0.455	0.413	0.391
D1	Papyrus	0.460	0.416	0.398
D2	Papyrus	0.458	0.395	0.378

*Series A cells had *Cyperus immensus* in phase 1. **Batch operation discontinued after 25 days and proceeded with PANPAPER's aerated pond wastewater on continuous flow.

Continuous flow

Phase 1: March 2004 - February 2005. Wetland cells series A-D were operated at a hydraulic loading rate (HLR) of between 4.1 – 4.9 cm/day while series E had HLR of 9.3 cm/day with wastewater from the final lagoon of PANPAPER Mills. The reason for adopting this operation mode was to improve the retention of suspended matter within the system and thereby enhance organic matter removal. Plant shoots were all harvested in December 2004 and the loading continued for a further two months till appreciable re-growth had been achieved. The *Typha* cells were then taken into a secondary treatment phase as described below.

Phase 2: April to August 2005. Wetland series A, C, D, and E were operated on continuous flow as subsurface systems while receiving wastewater from the final lagoon of PANPAPER Mills. The HLR for series A-D cells was in the range 4.9-5.7 cm/day while that for series E was 9.8 cm/day.

Secondary treatment: February to June 2005. The *Typha* cells were loaded on a continuous flow mode with wastewater from the second aerated lagoon of PANPAPER mills. The objective was to establish the level of pretreatment required in order to effectively treat pulp and paper mill wastewater in a wetland system.

Water budget

The water budget was determined for each cell using Equation 3.1. During batch loading wetland cell volumes were initially calculated from Equation 3.2. Later the volumes were determined by draining and refilling the cells with known amounts of wastewater. During the continuous flow operation the flows in and out of the wetland were measured twice daily by using the bucket and stopwatch method.

Precipitation was measured daily using a standard rain gauge installed at the site. Evapotranspiration was computed from Equation 1 in the case of continuous flow. For batch loading, however, the daily topping up volumes were used to calculate evapotranspiration. On-site measurements for temperature were taken using a portable Multi-parameter Water Quality Monitor; model 6820-10M-0, manufactured by YSI Incorporated, USA. Other seasonal climatological data including air temperature, relative humidity, and wind velocity were obtained from Nzoia and Bungoma meteorological stations, some 15 and 35 km west of the study site, respectively. The measured and computed data were used to determine the water balance for the cells. The budgets were used to calculate pollutant mass budgets and to determine removal efficiencies (Chapters 4 - 6).

Tracer Study

The tracer study was undertaken during continuous flow (phase 1) operation from 30th November to 7th December 2004. Lithium chloride was used as the tracer. For each wetland cell 1.5 g of lithium chloride (equivalent to 243.5 mg lithium) was dissolved in 1 litre of distilled water. The solution was mixed with influent in the inlet chamber and was carried into the system by the stream flow. The wetland outflow was sampled at three-hour intervals for eight days. However due to security reasons night-time sampling was less frequent with intervals of 9 – 12 hours from day 5 to day 8. The samples were preserved with nitric acid. Digests were analysed on a Varian model SpectrAA-10/20 atomic absorption spectrophotometer using the emission mode at 670.8 nm (APHA, 1995).

Results

Ambient conditions

Data available from Nzoia meteorological station indicated mean monthly day-time air temperatures in the range of 21.8°C to 26.8 °C with daily temperatures sometimes reaching 30

°C in the duration of the experiment. Relative humidity was low (38 – 67 %), while the mean monthly wind run reached a maximum of 160 km (1.9 m/s).

The temperature of the inflow wastewater was variable (22 °C to 30.5°C) but with no statistically significant difference (p>0.05) between the different phases of batch and continuous flow. The difference between wetland inflow and outflow water temperature was significant in batch operation but not in continuous flow. Mean outflow water temperatures were in the range 23.7±0.24 - 24.5±0.34°C. The larger free water surface flow cells had significantly lower water temperatures (mean of 22±0.25°C). There was no significant difference in temperature between SSF cells planted with different species. The SSF unplanted controls had similar mean temperatures as the planted ones except for the papyrus cells.

Water Balance

The wetland water balance was dominated by wastewater inflows and outflows (Tables 3.2 and 3.3). Rainfall was on average 6.8 – 8.9 mm/day and accounted for 12 to 14 % of inputs to individual SSF wetland cells (series A-D) and 7 % for the FWS cells (series E). Evapotranspiration rates were nearly double the rainfall and accounted for 22 - 32 % and 15 % respectively of outputs from the two system types. Rainfall was more variable over the experimental duration than ET. An example of this variation is depicted in Figure 3.3.

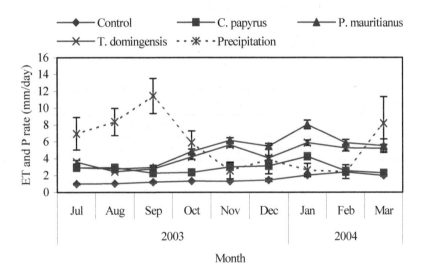

Fig. 3.3 Monthly evapotranspiration (ET) and Rainfall (P) in mm/day during batch operation phase 2 (3 days retention time).

Table 3.2 Water balance in continuous flow phase 1: March 2004 to February 2005 (n=752)

		Wetland cells*				
		Subsurface flow				Surface flow
		Control	Papyrus	*Phragmites*	*Typha*	*Typha*
Inputs						
Wastewater inflow	m^3/day	0.15	0.18	0.18	0.17	0.83
	se	0.002	0.002	0.002	0.002	0.003
Rainfall	m^3/day	0.025	0.025	0.025	0.025	0.061
	se	0.002	0.002	0.002	0.002	0.005
	mm/day	6.82	6.82	6.82	6.82	6.82
	se	0.52	0.52	0.52	0.52	0.52
Outputs						
Effluent outflow	m3/day	0.12	0.14	0.13	0.12	0.71
	se	0.003	0.002	0.003	0.003	0.006
ET	m^3/day	0.04	0.04	0.06	0.05	0.12
	se	0.001	0.002	0.002	0.001	0.005
	mm/day	10.25	10.03	15.36	12.22	13.62
	se	0.36	0.43	0.46	0.38	0.59

*Each cell type was replicated (pair) but data given is for mean of single cells

Table 3.3 Water balance in continuous flow phase2 (April – August 2005). All wetland cells including series E (*Typha2**) were operated as subsurface flow, n = 290

		Wetland cells				
		Control	Papyrus	*Phragmites*	*Typha*	*Typha2**
Inputs						
Wastewater inflow	m^3/day	0.21	0.21	0.21	0.18	0.87
	se	0.003	0.003	0.003	0.003	0.005
Rainfall	m^3/day	0.033	0.033	0.033	0.033	0.075
	se	0.058	0.058	0.058	0.058	0.058
	mm/day	8.87	8.87	8.87	8.87	8.87
	se	1.11	1.11	1.11	1.11	1.11
Outputs						
Effluent outflow	m3/day	0.19	0.19	0.17	0.14	0.77
	se	0.004	0.005	0.004	0.005	0.01
ET	m^3/day	0.03	0.03	0.05	0.06	0.14
	se	0.003	0.003	0.003	0.004	0.009
	mm/day	8.02	8.3	13.7	15.9	15.2
	se	0.79	0.69	0.92	0.99	1.06

Mean evapotranspiration rates during continuous flow were 8 – 16 mm/day which are 2 – 3 times the pan evaporation rates. Pan evaporation was in the range 4.4 –7.7 mm/day with an annual mean of 5.5 mm (Data from Nzoia Meteorological station). ET rates during batch operation were lower (up to 50 %) for SSF planted cells (Table 3.4, Figures 3.4 and 3.5) when compared to similar cells under continuous flow operation. The largest difference was noted for the unplanted wetland cells. These had a mean ET rate of 1.5 mm/day and 2.1 - 3.4 mm/day under the batch phases 2 and 3 respectively, nearly 5 times less than that under continuous flow.

Table 3.4 ET rates during wetland batch operation mode (n = 245, 240 and 75 for phases 1-3 respectively)

Wetland cell		ET rate mm/day		
		Phase 1 5-day HRT	Phase 2 3-day HRT	Phase 3 5-day HRT
A1*	Control	1,9±0.11	1,5±0.10	3,4±0.16
A2*	Control	2,2±0.06	1,5±0.06	2,1±0.12
B1**	*Typha domingensis*	5,5±0.20	4,5±0.17	4,7±0.54
B2**	*Typha domingensis*	4,9±0.19	4,1±0.14	4,7±0.54
C1	*Phragmites mauritianus*	4,1±0.13	5,2±0.22	7,7±0.55
C2	*Phragmites mauritianus*	3,3±0.11	4,8±0.18	7,2±0.55
D1	*Cyperus papyrus*	3,9±0.13	3,1±0.10	4.0±0.24
D2	*Cyperus papyrus*	2,7±0.08	2,7±0.09	4,4±0.27

* Planted with *Cyperus immensus* during phase 1. **Operated for 25 days only in phase 3

ET was expressed per shoot and per unit dry weight in order to provide a common "currency" for comparing rates for different species and operation modes. Mann-Whitney U test for independent samples was performed on ET data for wetland operations with similar shoot density and for different species. Comparison between continuous flow (phase 1) and batch loading with 3 days retention time (phase 2) revealed that the difference between the operational modes was statistically significant ($p < 0.05$).

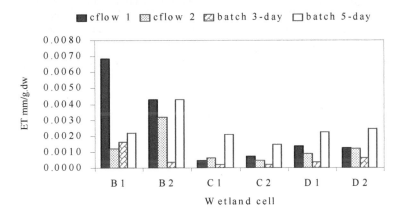

Fig. 3.4 Comparison of ET rates (mm per gram dry weight per day) for various wetland cells under different operation conditions – continuous flow phases1&2 (cflow) and batch phases 2&3. Cells B1 and B2 had *Typha*; cells C1 and C2 had *Phragmites* while cells D1 and D2 had papyrus.

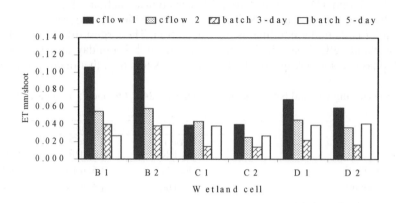

Fig. 3.5 Comparison of ET rates (mm per shoot per day) for various wetland cells at different operation conditions – continuous flow phases1&2 (cflow) and batch phases 2&3. Cells B1 and B2 had *Typha*; cells C1 and C2 *Phragmites* while cells D1 and D2 had papyrus.

The ET rates were different for different species and in the order *Typha* >> papyrus > *Phragmites*. When data for continuous flow phase 2 were compared to that of batch loading with 5 days retention time (phase 3) there was a significant difference only for ET expressed as mm/g.dw but not for ET expressed as mm/shoot. In the latter case, a significant difference was found only for *Typha*.

Wetland hydraulics

The tracer concentration-time response curves indicate that the tracer exited the wetland cells as an impulse (a sharp spike of concentration). Some examples of such curves are at Annex 3.1. There was a time delay (t_d) in tracer outflow from the various wetland cells. In general, the time delay was longer in the planted cells by up to 22 hours (Table 3.5). Peak tracer concentration was variable among the cells. The tracer was detectable as smaller peaks earlier and later than the main peaks. All t_d and peak times were longer than the nominal detention time (t_n). The latter was approximately 3 days when ET and rainfall are considered. Tracer recovery however, was poor and ranged from 7-75 %.

Table 3.5 Hydraulic characteristics for various SSF wetland cells for continuous flow (phase 1). ET and plant density are included for comparison.

Wetland cell	t_d hrs	t_p hrs	% Li recovery	Remarks
A1-Control	84	90	19	
A2-Control	84	87	7	Tracer detectable in effluent sampled at 63, 75 hours
B1-*Typha*	90	93	13	Tracer detectable in effluent sampled at 54, 78, 87 hours
B2-*Typha*	90	93	11	Tracer detectable in effluent sampled at 48, 63, 72, 170 hours
C1-*Phragmites*	106	118	75	Tracer detectable in effluent sampled at 69, 87, 170 hours
C2-*Phragmites*	81	84	16	Tracer detectable in effluent sampled at 75, 178 hours
D1-Papyrus	90	93	13	Tracer detectable in effluent sampled at 57, 72, 87, 174 and 178 hours
D2-Papyrus	84	87	27	Tracer detectable in effluent sampled at 57, 72, 146 hours

T_d = time delay before tracer outflow; t_p = time for peak tracer concentration

Discussion

Wetland hydrology

Warm wetland water temperatures prevailed throughout the experimental period and generally followed the air temperature. This is due to the intense solar radiation that prevails in the tropics all year round. Kadlec and Knight (1996) have reported the correlation between wetland water temperature and prevailing environmental conditions, in particular solar radiation, humidity and wind run that in turn influence evapotranspiration (ET). Okurut (2000) also made a similar observation for the water temperatures in a larger pilot scale constructed wetland receiving municipal wastewater in Jinja, Uganda. The consistently high wetland temperatures, according to (Kadlec 2005), are appropriate for microbial transformation processes that are typically dependent on water temperature.

High evapotranspiration rates (8 – 16 mm/day) obtained in this study are attributed to the prevailing intense solar radiation that resulted in warmer conditions (Kadlec and Knight, 1996; Kadlec, 2005). Under arid conditions ET losses could be exceedingly high. Ranieri (2003) reported ET losses of up to 40 mm/day for the hottest summer days in semi arid conditions of southern Italy. There is a paucity of data on ET rates from constructed wetlands in the tropics and more so in east Africa. Okurut (2000) reported mean ET rates of 6.1 and 5.6 mm/day for papyrus and *Phragmites mauritianus* cells respectively. The mean ET rate (13.6 mm/day) for the FWS wetland with *Typha domingensis* in this study is about twice the Jinja rates. The difference between the findings of this study and that of Okurut (2000) may be attributed to differences in the prevailing environmental factors especially relative humidity of the sites. Webuye experiences lower humidity than Jinja. The latter is located in close proximity to a large body of water, Lake Victoria, and a natural wetland. Kadlec and Knight (1996) alluded to the "clothesline effect" on small wetlands, which may be partly responsible for the higher ET rates found in this study.

Evapotranspiration forms a significant fraction of the hydraulic loading (Kadlec and Knight, 1996). In this study, ET was about 30 % of the hydraulic loading for planted cells and 20 % for the unplanted cells. ET from the wetland accounted for 22 - 32 % and 15 % of the outputs from the SSF and FWS cells respectively, and was twice the rainfall depth. Okurut (2000) obtained similar proportions of water loss (15-20 %) from the Jinja FWS constructed wetland.

Influence of plants and operation mode
The evapotranspiration rates for wetland cells with various plant species were different (Figures 3.4 and 3.5). *Typha*-planted cells had the highest rates in nearly all operation modes except during batch phase 3 (5 days retention). For the same plant species, ET rates seemed to depend on shoot density and plant biomass.

ET rates were higher during continuous flow than batch operation mode provided shoot density and biomass was similar. This was the case when data for continuous flow phase 1 and batch phase 2 operations were compared. The situation was however different when continuous flow phase 2 and batch phase 3 data were compared. The reason for the difference is that although shoot density was similar during both operation modes, plant shoots were at a more advanced age and hence had higher biomass during the continuous flow period (Chapter 4). Higher ET rates during the continuous flow operation are due to the constant water level in the wetland bed. During batch operation on the other hand, water levels were declining during the day until replacement (topping up) the following morning. For the control, unplanted ponds the large difference in ET rates between the different operation modes may have been caused by differences in water temperature. Evaporation through the gravel surface depends mainly on the prevailing water temperature to provide the kinetic energy. During batch operation the difference between inflow (loaded water temperature) and *in-situ* water temperature was found to be significant. During the continuous flow operation on the other hand, there was no significant difference between inflow and outflow water temperature.

Implication of ET
The mean evapotranspiration rate exceeded rainfall and is an important component of outputs (water loss) in the water budget of the wetland system. ET was found to reduce the water flow rate in the wetland system, increasing the nominal hydraulic detention time by nearly a day. The actual hydraulic detention time, however, is a function of wastewater flow paths and extent of wastewater interaction with the bed media and vegetation, and may be shorter (see below).

ET losses in batch loading were compensated for daily by topping up with presettled river water. In a full-scale constructed wetland however, topping up would complicate wetland operation and may increase costs. If topping up were not done ET losses in a five-day batch cycle for instance, may be more than 25 mm/day, which would increase pollutant concentrations in the outflow.

Flow characteristics

The wetland tracer flow peaks emerged, after a time delay, as a sharp spike. This is characteristic of gravel beds with horizontal sub-surface flow. The time delay is caused by severe decline in (both planted and unplanted gravel beds) hydraulic conductivity in the front (inlet) end of the wetland bed due to deposition of sediments and detritus (Kadlec and Knight, 1996). In this study, the time delay for planted wetland cells was 81 – 106 hours and in the unplanted cells 84 hours. The longer delay for planted cells may be due to the presence of plant roots and rhizomes at the inlet and to higher ET losses, especially in the *Typha* and *Phragmites* cells. The sharp tracer peaks indicate that there was little mixing and the water predominantly moved as a plug. The additional smaller peaks, according to Kadlec and Knight (1996) are indicative of smaller flow channels that do not mix with the main one.

Both the time delay (t_d) and the time for emergence of peak tracer (t_p) concentration were higher, by up to a factor of two, than the nominal detention time. This presents a peculiar but important finding in that with a conservative tracer, such as lithium chloride is assumed to be, the nominal detention time, which assumes 100 % hydraulic efficiency, is usually higher than both the tracer t_d and t_p (Kadlec and Knight, 1996; Persson *et al.*, 1999). There are two possible reasons for this observation. Firstly, the lithium tracer may have been partly adsorbed in the gravel bed especially at the inlet where there is organic matter accumulation. The low recovery of the tracer indicates that adsorption might have occurred and retarded its transport through the substrate medium. Mojid and Vereecken (2005) described lithium as a non-linearly sorbing solute in a study of contaminant plume transport in a ground water aquifer. Secondly, there may have been re-circulation or dead zones in the gravel bed into which water moves and stays for some time before getting back into the main flow channels (Kadlec and Knight, 1996; Kadlec and Watson, 1993). The smaller tracer peaks that emerged after the main one in nearly all the cells support this behaviour. Low tracer recovery, however, may also be partly due to the long (grab) sampling interval especially in the night, 9 – 12 hours. The large difference in the magnitude of tracer peaks is indicative of the variation in channelling (preferential flow paths) in the gravel beds. This in turn is due to differences in below ground biomass (root and rhizome) and non-uniformity in the gravel packing within the bed. Dumping and filling of gravel in SSF cells leads to diverse arrangement of gravel particles within the bed.

Conclusion

Evapotranspiration is an important component of outputs in the water budget of the wetland system. It should therefore be an integral part in wetland design in the tropics. ET rates are different for different aquatic plant species. Therefore ET may be an important selection criterion for plant species to be used in a constructed wetland used for wastewater treatment especially in drier climate.

ET rates are higher in continuous flow compared to batch loading with daily compensation. However, in a full-scale wetland without topping up, the likelihood of high ET losses in batch operation and the resulting pollutant concentration makes batch operation undesirable.

It was not possible to deduce the actual retention time and other hydraulic parameters (efficiency and number of "tanks in series") as there was no discernable tracer concentration curve for all wetland cells. Although ET causes an increase in nominal retention time it may be predicted that due to the presence of dead or recirculation zones and short-circuiting the actual retention time would be close to but slightly lower than the nominal. In addition, it may be deduced from the sharp spikes that some measure of plug flow did take place albeit following several micro channels.

For pulp and paper mill wastewater, which has high organic matter content, the tracer study should be conducted with a different tracer. Alternatively, lithium chloride may still be used but with continuous feed instead of pulse feed, as was the case in this study. The continuous dosing would ensure that the lithium adsorption capacity of the bed is exceeded and tracer breakthrough is achieved. Determination of the adsorption capacity of accumulated sludge is necessary in determining the tracer dosage into the wetland. Due to logistical constraints it was not possible to repeat the tracer study.

References

Abira M. A, Ngirigacha H. W. and van Bruggen J.J.A. 2003. Preliminary investigation of the potential of four tropical emergent macrophytes for treatment of pre-treated pulp and papermill wastewater in Kenya. *Water Science and Technology* 48 (5): 223 - 231.

Abira, M.A., van Bruggen, J.J.A. and Denny, P. 2005. Potential of a tropical subsurface constructed wetland to remove phenol from pre-treated pulp and papermill wastewater. *Water Science and Technology* 51 (9): 173 - 176.

APHA, 1995. *Standard methods for the analysis of water and wastewater*, 19th edition. American Public Health Association, Washington DC.

Boyd, J., McDonald, O. Hatcher, D., Portier, R.J., and Conway, R.B. 1993. Interfacing constructed wetlands with traditional wastewater biotreatment systems. In: Moshiri, G.A. (ed.). *Constructed wetlands for water quality improvement*. CRC Press, Lewis Publishers, Michigan, pp 453 - 460.

Cooper, P.F., Job, G.D., Green, M.B., and Shuttes, R.B.E. 1996. *Reed beds and constructed wetlands for wastewater treatment*. WRc plc. Swindon, Wiltshire, U.K., 184, pp.

Hammer, D.A., Pullin, B.P., Mc Murry, D.K. and Lee, J.W. 1993. Testing color removal from pulp mill wastewaters with constructed wetlands. In: Moshiri, G.A. (ed.). *Constructed wetlands for water quality improvement*. CRC Press Lewis Publishers, Michigan, pp 5 – 19.

JICA/GOK, 1992. *National Water Master Plan, Data Book: Hydrological data*. Japan International Cooperation Agency and Government of Kenya, Ministry of Water Development.

Kadlec, R.H. 2000. The inadequacy of first-order treatment wetland models. *Ecological Engineering* 15: 105 - 119.

Kadlec, R.H. 2005. Wetland to pond gradients. *Water Science and Technology* 51 (9): 291 - 298.

Kadlec, R.H., 1989. Hydrologic factors in wetland wastewater treatment. In: Hammer, D.A (ed.), *Constructed wetlands for wastewater treatment: municipal, industrial and agricultural*. Lewis Publishers, Chelsea, Michigan, U.S.A., pp. 21 - 40.

Kadlec, R.H., Bastiaens, W. and Urban, D.T., 1993. Hydrological design of free water surface treatment wetlands. In: Moshiri, G. A. (ed.). *Constructed wetlands for water quality improvement*. CRC Press, Lewis publishers, Michigan, pp 77 – 86.

Kadlec, R.H. and Knight, R.L. 1996. *Treatment wetlands*. CRC Press Inc., Boca Raton, Florida, 893 pp.

Kadlec, R.H. and Watson, J.T., 1993. Hydraulics and solids accumulation in a gravel bed treatment wetland. In Moshiri, G.A. (ed). *Constructed wetlands for water quality improvement*, CRC Press, Lewis Publishers, Boca Raton, pp 227-235.

Metcalf and Eddy, Inc., (1991). *Wastewater engineering: treatment disposal and re-use*. Third edition. Revised by G. Tchobanoglous and F.L. Burton, McGraw-Hill, New York, USA, 1334 pp

Meuleman, A.F.M. (1999). Performance of treatment wetlands. PhD. Dissertation, Utrecht University, The Netherlands.

Mojid, M.A. and Vereecken, H. 2005. On the physical meaning of retardation factor and velocity of a non-linearly sorbing solute. *Journal of Hydrology* 302 : 127 - 136.

Moore, J.A., Skarda, S.M. and Sherwood, R. 1994. Wetland treatment of pulp and paper mill wastewater. *Water Science and Technology* 29 (4): 241 - 247.

Okurut, T. O. 2000. A pilot study on municipal wastewater treatment using a constructed wetland in Uganda. PhD dissertation, Balkema publishers, Rotterdam, The Netherlands

Persson, J., Somes, N.L.G. and Wong, T.H.F. 1999. Hydraulic efficiency of constructed wetlands and ponds. *Water Science and Technology* 40 (3): 291 - 300.

Ranieri, E. 2003. Hydraulics of sub-surface flow constructed wetlands in semi-arid climate conditions. *Water Science and Technology* 47 (7-8): 49 - 55

Reed, S.C., Crites, R.W., and Middlebrooks, E.J. 1995. *Natural systems for wastewater management and treatment*. McGraw-Hill, New York.

Tchobanoglous, G. 1993. Constructed wetlands and aquatic plant systems: research, design, operational and monitoring issues. In: Moshiri, G. A. (ed.) *Constructed wetlands for water quality improvement*.CRC Press, Lewis publishers, Boca Raton, pp 23 – 34.

Thut, R.N., 1993. Feasibility of treating pulp mill effluent with a constructed wetland. In: Moshiri, G.A. (ed.). *Constructed wetlands for water quality improvement*. CRC Press, Lewis Publishers, Boca Raton, pp 441-447.

US EPA, 1988. *Design manual: Constructed wetlands and plant systems for municipal wastewater treatment*. U.S. Environmental Protection Agency, Cincinnati.

Watson, J.T., Reed, S.C., Kadlec, R.H., Knight, R.L., and Whitehouse, A.E. 1989. Performance expectations and loading rates for constructed wetlands. In: Hammer D.A. (ed.), *Constructed wetlands for wastewater treatment: municipal, industrial and agricultural*. Lewis Publishers, Chelsea, Michigan, U.S.A., pp. 319-351.

WPCF (Water Pollution Control Federation) 1990. *Natural systems for wastewater treatment*. Manual of Practice FD-16. Alexandria, VA. Washington, DC.

Young, D.F., Munson, B.R., and Okiishi, T.H. 1997. *A brief introduction to fluid mechanics*. John Wiley, New York.

Annex

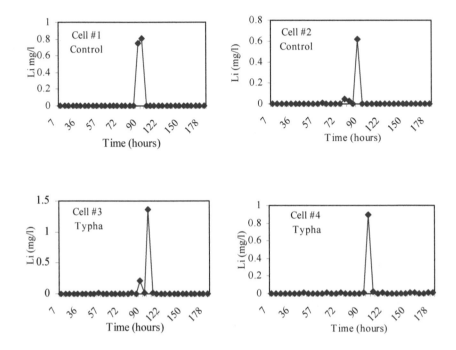

Annex 3.1 Tracer concentration-time response curves for cells series A (control: cell no 1 and 2) and series B (*Typha*: cell numbers 3 and 4)

Chapter four

Plant growth and nutrient removal efficacy

Publication based on this chapter:

Abira, M. A., Ngirigacha, H. W. and van Bruggen J.J.A. 2003. Preliminary investigation of the potential of four tropical emergent macrophytes for treatment of pre-treated pulp and papermill wastewater in Kenya. *Water Science and Technology* 48 (5): 223 – 231.

Plant growth and nutrient removal efficacy

Abstract
The growth of four tropical emergent aquatic macrophytes and their efficacy in removing nutrients (nitrogen and phosphorus) from pre-treated pulp and paper mill wastewater was studied in a subsurface flow constructed wetland in western Kenya. The wetland consisted of four pairs of cells planted with *Cyperus immensus*, *Typha domingensis*, *Phragmites mauritianus* and *Cyperus papyrus*. *C. immensus* was removed after eight months and the cells left as unplanted controls. Both batch and continuous flow modes of operation were employed using the final effluent from Pan African Paper Mills treatment ponds. The wetland inflow wastewater had mean total nitrogen and phosphorus concentrations of about 3 mg/l and 0.7 mg/l respectively giving a low N:P ratio. Biomass (as dry weight), tissue nutrient concentration and uptake rates were determined at different stages of growth. The wetland efficacy in removing nutrients was determined based on mass flows.

Plant tissue nutrient concentrations were lower than in healthy natural wetland plants. Nitrogen concentrations based on dry weights in *Phragmites*, *Typha* and papyrus were 9.2±0.7, 7.4±0.5, and 6.1±0.2 mg/g, respectively while phosphorus concentrations were 1.7±0.12, 1.9±0.11, 1.6±0.14 mg/g, respectively. *Typha* and *Phragmites* had satisfactory above ground biomass production (10896 g/m^2 and 3015 g/m^2 respectively) when compared to natural wetlands. The growth of papyrus was sub-optimal with aerial biomass of 3075 g/m^2. Plant vitality and growth was lower during batch mode wetland operation. Papyrus appears to be sensitive to lack of water and was affected by periods of dryness in the load-drain batch operation, recovering at the commencement of continuous flow.

Mean removal efficiency for total nitrogen was in the range 49 - 75 % for planted cells and 42 - 49% for unplanted ones in continuous flow. Series E *Typha* cells with nearly double the hydraulic loading rate of other cells achieved 57 % removal. During batch operation modes nitrogen removal efficiencies were in the range 32 - 68 % for planted cells (excluding *C. immensus*) and 29 - 35 % for the unplanted. Removal efficiency of up to 25 % was attributed to the presence of plants. Phosphorus removal was in the range of 30-60 % in planted cells and -4 – 38 % in unplanted ones in continuous flow compared to 10 – 58 % in planted and -9 - 12 % in unplanted cells under batch operation. *Phragmites, Typha* and papyrus (in that order of preference) were all found to be suitable for the removal of N and P from pulp and paper mill effluents. A mixture of all three species might provide the best long-term efficacy appropriate for the changing and varying environmental conditions.

Key words
Cyperus immensus, Typha domingensis, Phragmites mauritianus, Cyperus papyrus, macropyhtes, nutrient removal, nutrient budgets,

Introduction

In this chapter we investigate the role of emergent aquatic macrophytes in a constructed wetland pilot project for the treatment of industrial wastewater from a pulp and paper mill in Kenya, in relation to nutrient removal. Pan African Paper Mills (E.A.) Limited (PANPAPER) was selected as a site for the constructed wetland tertiary industrial effluent treatment pilot project under the Lake Victoria Environmental Management Project (LVEMP). The factory is an integrated pulp and paper mill with an average annual production of about 120,000 tonnes. Large amounts of wastewater are produced (32,000 – 37,000 m^3 per day) that are discharged into the Nzoia River after primary clarification and secondary treatment in aerated lagoons. The effluent seldom complies with national regulations for biochemical oxygen demand (BOD), chemical oxygen demand (COD), total suspended solids (TSS) and phenols. Some characteristics of the final treated effluent from the lagoons are presented in Table 4.1. A detailed description is given in Chapter 2. The nitrogen (2.56±0.40 mg/l) and phosphorus (0.76±0.05 mg/l) concentrations in the final effluent that were fed to our study constructed wetland are similar to those found in natural wetlands such as the fringing wetlands of Lake Naivasha, Kenya (2.51 mg/l total N and 0.15 mg/l total P - Muthuri and Jones, 1997), and the Nawandigi and Lubingi wetlands on the northern shores of lake Victoria, in Uganda (ca 2.5 mg/l total N and 1.0 mg/l total P – Kipkemboi *et al.*, 2001) but an order of magnitude less than concentrations found in either domestic sewage (mean 60 mg/l ammonium and 1.4 - 6.5 mg/l P   (Okurut, 2000); 15 - 29 mg/l total kjeldahl nitrogen   (Koottatep and Polprasert, 1997) or in natural wetlands receiving domestic sewage such as the Nakivubo and Namiiro wetlands fringing Lake Victoria in Uganda with up to 20 mg/l total N and 1.4 mg/l total P (Kipkemboi *et al.,* 2001; Kansiime *et al.*, 2003). The aim of the pilot project at PANPAPER was to investigate the potential of the constructed wetland for treating the effluent of the lagoons before discharge into the Nzoia River, and therefore prevent the river from further degradation.

Table 4.1 characteristics of the final effluent from PANPAPER Mills wastewater treatment ponds. A portion of the flow was diverted into the pilot study constructed wetland. Data were collected from September to December 2002 (n = 6) and between July 2003 and June 2005 (n = 4).

Parameter/unit	Mean	se
pH	7.92	0.12
Dissolved oxygen, mg/l	0.22	0.019
Total dissolved solids, mg/l	900	191
Phenols, mg/l	0.64	0.09
BOD mgO$_2$/l	45	3
COD mgO$_2$/l	394	340
Total suspended solids, mg/l	52	6
Total nitrogen, mg/l	2.56	0.40
Total phosphorus, mg/l	0.76	0.05

Macrophytes in constructed wetlands

The presence or absence of wetland plants is one of the characteristics often used to define the boundary of wetlands and thus is an inherent property of wetlands, including constructed wetlands (Brix, 1997). Wetlands have individual and group characteristics related to plant species present and their adaptation to specific hydrologic, nutrient and substrate conditions

(Gunstenspergen *et al.*, 1989). Because of this a variety of plant species are used in constructed wetland systems. The species may be submerged, emergent or floating types. In this study only emergent species were used.

Typical emergent aquatic macrophytes used in constructed wetlands in temperate countries and in Australia are the common reed (*Phragmites australis*), cattail (*Typha latifolia*), and bulrush (*Scirpus lacustis*) (Brix, 1993). In the wetlands of tropical Africa the most important genera of emergent freshwater macrophytes include *Cyperus, Phragmites, Panicum, Scirpus* and *Typha* (Thompson, 1985). *Typha domingensis* is widespread in tropical and subtropical zones. It is a hardy plant and can persist for several years in the absence of a flooding regime. It is more drought and salt tolerant than *Phragmites. Typha domingensis* has colonized a marshy area in the vicinity of the wastewater treatment ponds of PANPAPER Mills. *Typha*'s normal above ground production is in the range 1.5 - 2.5 kg dw/m^2. *Phragmites mauritianus* has an exclusively tropical distribution in Africa along river banks and can grow to 5 metres or more with above ground biomass production of 2 - 5 kg/m^2 dry weight (Thompson, 1985). In a pilot study constructed wetland for treatment of municipal wastewater in Jinja, Uganda, standing biomass productivity reached a maximum of 104 tons ha^{-1} (ca 10 kg/m^2) during the exponential growth stage (Okurut, 2000).

The *Cyperaceae* are widely distributed in east and central Africa. *Cyperus papyrus* is the largest sedge in the world with an average height of 4 - 5 m but can grow up to 9 m and has a highly branched umbel at least 50 cm in diameter (Thompson, 1985; Denny, 1993). Okurut (2000) reported standing biomass productivity of up to 108 ton ha^{-1} for papyrus in the sewage-fed Jinja wetland. *Cyperus immensus* is widespread in Kenya being found in flood plains, and on the outer edges of wetlands dominated by papyrus. These tropical wetland species are indigenous to Kenya and are found in several areas of the Lake Victoria basin. However, their use in constructed wetlands in Kenya has not been exploited until recently. For example, *Typha* and papyrus are among the plants used in the Carnivore-Splash constructed wetland in Nairobi, Kenya that treats combined wastewater from a restaurant and a swimming pool resort with recreational facilities (Nyakang'o and van Bruggen, 1999). The Oserian constructed wetland that treats combined sewage, pack house waste and bucket (chemical) rinsate from a large flower farm in Naivasha, Kenya, is mainly planted with *Scirpus* species (Bett, 2006).

Role of Emergent macrophytes

Brix (1997) succinctly summarised the role of macrophytes growing in constructed wetlands in relation to the treatment processes as: physical effects (e.g. erosion control, filtration, provision of surface area for attached organisms); and metabolism of the macrophytes (nutrient uptake, oxygen release). Emergent macrophytes take up ions (including nutrients) through their roots and translocate them into their shoots (Denny, 1987). Later during growth, some nutrients are relocated to rhizomes, seeds and other storage tissues while some are lost into the surrounding water or sediment through tissues senescence or decay.

Macrophyte growth on pulp and paper mill wastewater

In previous work with pulp mill wastewater in the United States of America, aquatic macrophytes used varied widely by family/genera and included *Cyperaceae (Scirpus), Poaceae (Panicum, Phragmites* and *Spartina), and Typhaceae (Typha)*. They were selected on the basis of local availability, growth rate during periods of low flow in the receiving water body (Thut, 1989), and product use (Thut, 1993). Plant growth was reported to be prolific in pulp mill wastewater with above ground tissue reaching 4.4 kg/m^2 in a year (Thut, 1993). In the latter study, total nitrogen and phosphorus concentrations in the inflow wastewater were low (3 mg/l and 0.6 mg/l respectively). The studies did not report on any potential nutrient deficiency effects.

Preliminary experiments in bucket mesocosms at the PANPAPER Mills in Kenya (Abira *et al.*, 2003) tested a range of indigenous water plants including *Cyperus papyrus, Phragmites mauritianus, Cyperus immensus* and *Typha domingensis* to establish their tolerance to primary and secondary treated wastewater as well as the ability to improve the quality of effluent. In these experiments the N and P concentrations in the influent water were generally low. In the primary treated wastewater the concentration of orthophosphate-P, nitrate-nitrogen and ammonium nitrogen averaged 0.047±0.014, 4.94± 1.88 and 2.51± 0.29 mg/l respectively while in the secondary treated wastewater the corresponding concentrations were 0.047±0.014, 1.81±0.47 and 4.2±0.47 mg/l respectively. All the macrophytes survived in all treatments but appeared greener and healthier in treatments with a retention time of 5 days compared to 10 days. Plant nutrient storage was higher at 5 days. Plant nitrogen and phosphorus content, based on dry weight, was lower at the end of the experiment than at the beginning in all treatments for all species. From these preliminary experiments it was decided that the same species could be used in the current pilot project (Chapter 3).

Nutrient removal efficiency and processes

Nitrogen and phosphorus removal in wetlands is variable and depends to a large extent on the systems loading and the ambient conditions such as temperature, pH, and dissolved oxygen concentration (Kadlec and Knight, 1996). Removal efficiencies in the range of 44 - 90 % for nitrogen and 34 - 88 % for phosphorus have been reported for some tropical constructed wetlands receiving sewage (Senzia *et al.*, 2002; Okurut, 2000; Nyakang'o and van Bruggen, 1999; Juwakar *et al.*, 1995).

Nitrogen transformations in wetland ecosystems are very complex. The mass balance is determined by inflow wastewater concentrations and environmental factors that affect various processes that continually transform inorganic to organic forms and *vice-versa* (Vymazal, 2001). The processes include ammonia volatilisation, nitrification, denitrification, nitrogen fixation, plant uptake and decay, microbial uptake and decay, mineralisation, nitrate reduction to ammonium, fragmentation, adsorption, desorption, burial and leaching. In most treatment wetlands the sequential nitrification-denitrificatiom plays a major role compared to plant uptake and other processes (Kadlec and Knight, 1996; Reddy *et al.*, 1989).

Biomass harvesting, at least in temperate climate systems, reportedly does not remove significant amounts of nutrients in wetlands used as secondary treatment systems (Brix, 1994). In wetlands receiving low loads of nitrogen, however, plant uptake plays a significant role. Toet (2005) reported up to 11 % nitrogen removal in low-loaded *Phragmites* and *Typha*-planted ditches used for polishing pre-treated sewage in The Netherlands. In contrast, biomass harvesting in constructed wetlands located in tropical climatic regions where the growing season is not limited by temperature and seasonality, can remove significant amounts of nutrients especially when the harvest is done in the exponential growth stage. In a study on a constructed wetland receiving pre-settled municipal sewage in Uganda, Okurut (2001) found that plant uptake contributed to 39 % and 86 % of the nitrogen; and 33 % and 61 % of the phosphorus removal by *Cyperus papyrus* and *Phragmites mauritianus* respectively. In Thailand, Koottatep and Polpasert (1997) found that nitrogen uptake by *Typha augustifolia* accounted for about 50% of the total mass input in a free water surface flow wetland fed with primary treated sewage. Harvesting of plants at 8-week intervals yielded higher nitrogen uptake amounting to 66 - 71% of input to the wetland. Tanner (1996) attributed mean removals of up to 30 % of inflow nitrogen and phosphorus to uptake by diverse plant species (including *Baumea, Phragmites* and *Zizania*) in a constructed wetland receiving primary treated dairy wastewater in New Zealand. Greenway (1997) and Kipkemboi *et al* (2002) concluded that there is a good potential for nutrient removal via plant uptake. The former study determined tissue nutrient content of various macrophyte species including *Typha domingensis* and *Cyperus involucratus* in several pilot scale wetlands in tropical, subtropical and arid locations of Australia while the latter determined biomass production of *Miscanthidium violaceum* and *Cyperus papyrus* in both natural (receiving and not receiving sewage) and laboratory-scale constructed wetlands in Uganda. In all the studies plant nutrient tissue content and/or uptake rates were higher in wetlands receiving nutrient-rich wastewater compared to ones that were not.

Constructed and natural wetlands are capable of absorbing new phosphorus loadings and depending on the prevailing conditions can provide both short-term and long-term storage of the nutrient (Kadlec *et al.*, 2000). Phosphorus transformation in wetlands include adsorption, desorption, precipitation, dissolution, plant and microbial uptake, leaching, mineralisation, sedimentation and burial (Vymazal, 2001). Sustainable removal processes include sediment accretion and residual detritus from macrophytes. As with nitrogen, plant uptake and subsequent harvesting can play a significant role in low-loaded systems.

Study objectives

The objectives of the study were to establish the growth characteristics of the four emergent macrophytes in pulp and paper mill wastewater under varying operating conditions, to assess nutrient removal efficiency and to determine mass flows for maintaining healthy plant communities.

Materials and methods

Macrophyte growth

The study wetland construction, set-up, planting and operation are described in Chapter 3. A summary of the different phases of operation is given in Table 4.2. Quadrants were selected and defined at either ends of the planted beds and in the middle. Shoot counts and height measurement were made periodically. Maximum leaf number was monitored for *Phragmites mauritianus, Cyperus immensus* and *Typha domingensis*. Culm and stem girth were measured for *Cyperus papyrus* and *Phragmites* respectively while umbel length/diameter was monitored for papyrus. At 800 days (10th December 2004) from the start of experimentation all wetland cells were cut and cleared of above ground plant material and shoot re-growth was monitored for a further 240 days.

Wastewater nutrient concentrations

Nitrogen and phosphorus (total and dissolved) concentrations in wetland inflow and outflow waters were sampled bi-weekly together with the other parameters (Chapters 5 and 6) for analysis in the laboratory. The results obtained together with flow data (Chapter 3) were used to evaluate the system's efficiency in removing nutrients.

Table 4.2 Summary of experimental design and operation with four pairs of subsurface flow cells. Details are given in Chapter 3. Wetland cells were irrigated with river water initially followed by a mixture of treated effluent and river water during acclimation. During the study all cells were loaded with the final effluent from PANPAPER Mills treatment ponds except as indicated below.

	Operation/ Growth Stage	Duration (days)	2002						2003												2004												2005							
			Jul	Aug	Sep	Oct	Nov	Dec	Jan	Feb	Mar	Apr	May	Jun	Jul	Aug	Sep	Oct	Nov	Dec	Jan	Feb	Mar	Apr	May	Jun	Jul	Aug	Sep	Oct	Nov	Dec	Jan	Feb	Mar	Apr	May	Jun	Jul	Aug
1	Planting	1	■																																					
2	Acclimation	45				■	■																																	
3	Batch Phase 1 (5 - Day)	242											■	■	■	■	■	■	■	■																				
4	Batch Phase 2 (3 - Day)	241													■	■	■	■	■	■	■	■																		
5	Continuous Flow Phase 1	325																			■	■	■	■	■	■	■	■	■	■	■	■								
6	Batch Phase 3 (5 - Day)	75																															■	■	■					
7	Continuous Flow Phase 2	135																																		■	■	■	■	■
8	Continuous Flow Phase 2 (*Typha* Cells)	60																																			■	■		

Typha cells were loaded with wastewater from the second aerated pond of PANPAPER Mills from February to June 2005. Exponential growth stage = February to May 2003 (Crop 1) and February to April 2005 (Crop 2); steady growth stage = March to July 2004 (Crop 1). All shoots harvested at 800 days from start.

Biomass and nutrient determination

Biomass (as fresh and dry weight) was determined at the beginning and end of each experimental mode of operation. Plant samples for biomass determination were initially (November 2002) taken by uprooting whole plants and segregating into different parts (root and rhizome, shoot, culm and umbel as applicable). Later, as both above and below ground biomass increased, harvesting of below ground parts was not feasible, as it would have disrupted ongoing experiments. Samples for fresh weight were washed thoroughly under running tap water, rinsed with distilled water and blotted dry before weighing. For dry weight determination the cleaned samples were oven dried at 70˚C for three to five days until constant weight.

At the end of the study the above ground biomass was harvested. For below ground biomass a section in the middle of each cell was cut through the gravel substrate across the cell width, down to the cell bottom (dimensions measuring 1.2 m x 0.50 m x 0.3 m). Rhizomes were segregated from roots and both fresh and dry weights were determined. The ratio of fresh weight to dry weight was used to estimate the total below ground biomass.

Relative growth rate was calculated according to Kvet and Westlake (1998) using the following expression:

$$G = \frac{(\ln W_{final} - \ln W_{o})}{\delta t}$$
Equation 4.1

Where:

G	=	Unit change in biomass per unit time (d^{-1})
W_{final}	=	Final mean plant dry weight in grams
W_{o}	=	Plant mean weight at time zero in grams
δt	=	Duration of monitoring in days

Samples for nutrient content (nitrogen and phosphorus) determination were oven dried at 70° C to constant weight and ground to a fine mesh size. Replicate sub-samples were digested using selenium acid mixture as described by Novozamsky et al. (1983). The digests were analysed using a spectrophotometer according to the *Standard Methods* (APHA, 1995).

Nutrient mass flows

Nitrogen and phosphorus flows were determined from the mass inputs (inflow wastewater), outputs (outflow wastewater) and plant uptake of the wetland cells for the periods February to May 2003 and March to July 2004 corresponding to the exponential and steady growth stages of plant shoots. The periods correspond to months 3 - 6 and 16 - 20 of the wetland operation respectively. Nutrient accumulation and/or release in sediments prior to and during the periods under consideration were not determined. Input from rainfall and output by evapotranspiration were assumed to be negligible.

Data analysis

Data were analysed using Microsoft Excel analysis toolpak. Means were compared using SPSS statistical software. Comparisons of data obtained at the beginning and at the end of each operation phase were made using the two-sample Kolmogorov-Smirnov test.

Results

Plant growth characteristics

General appearance
During the continuous flow operation plant shoots appeared more lush green than the batch-loaded systems, especially for papyrus and *Phragmites*. Emerging juvenile shoots had thicker stems. In general plants appeared more lush-green and healthier in the re-growth phase than at the first crop. Plants were taller and had higher vitality at the inlet end than at the outlet ends of cells in continuous flow. Some of the features of plant growth are presented in Table 4.3.

Table 4.3 Features of wetland plant growth during continuous flow phase 1 (crop I) and phase 2 (crop II). *Cyperus immensus* was lush green but frequent attacks by vermin monkeys led to patchy establishment and is not included here.

Species	% Flowering	Umbel length, cm or leaf no	Umbel diameter cm	Stem or Culm girth, cm	Senescing or dead, (number of shoots)	Remarks
Crop I						Initial
Typha	37	10-14			5	
Phragmites	0	22-28		7.2 - 9	6	Leaf yellowing and fall in September-October 2003
Papyrus	6 - flowering shoots <1.5 m tall	Max 55 (umbel)	50-60	9 - 11	60 All buds, juveniles died by month 6	Plants had a yellowish appearance
Crop II						Re-growth
Typha	13	12-15			3	
Phragmites	2	24-30		5 - 8	0	
Papyrus	11 - most flowering shoots >2.5 m tall	50-60 (umbel)	60-80	7 - 11	15	Plants had a lush green appearance

Shoot height
Increase in shoot height was more rapid in *Phragmites* and *Typha* than in papyrus with plants attaining mean maximum heights of 4.9 m, 3.3 m, and 2.8 m respectively for the three species within 12 months (by December 2003). The heights were maintained over the following 10 months (up to December 2004). The mean maximum height attained by *C. immensus* was 1.1 m in 8 months. This species was removed at this time and considered unsuitable for further experimentation. During the shoot re-growth phase heights attained in 7 months were 3.8 m, 3.6 m, and 2.2 m for *Phragmites, Typha*, and papyrus. Mean maximum heights attained by the steady growth stage (12 months) were 4.8, 4.4 and 3.4 m respectively.

Shoot density
The shoot density (Figure 4.1) in all wetland cells (*Typha, Phragmites* and papyrus) increased rapidly from the initial 7.2 shoots /m^2 to 50, 84, and 65 shoots /m^2 respectively in 8 months (Batch phase 1). The increase was slower in the subsequent 8 months (Batch phase 2) with both papyrus and *Typha* having attained their maxima (68 and 53 shoots/m^2 respectively). *Phragmites* continued to increase in numbers reaching 150 shoots/m^2 in the same period.

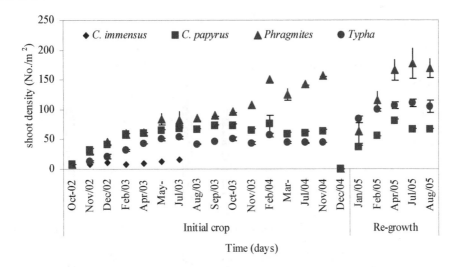

Fig. 4.1 variations in shoot density over the experimental period. Plant shoots were all harvested in December 2004 giving way to fresh growth (January – August 2005).

Biomass productivity and growth rates

Standing biomass

Plant re-growth was prolific in the period following the harvesting of the shoots. Within four months the accumulated biomass for papyrus, *Typha* and *Phragmites* was similar to the maxima attained in 22 months of growth in the initial crop. The variation in above ground (shoot) aerial biomass yield (as dry weight) is depicted in Figure 4.2. In the initial crop, *Typha* attained its maximum biomass (3015 g/m^2) earlier (July 2004) than *Phragmites* and papyrus (10896 g/m^2 and 3075 g/m^2 respectively in December 2004). *Phragmites* had the highest overall biomass production rate (21 g/m^2.day) compared to *Typha* (15.4 g/m^2.day) and papyrus (14.3 g/m^2.day). The same trend was maintained during the shoot re-growth period. The biomass production rates were 81 g/m^2.day, 28 g/m^2.day and 25 g/m^2.day respectively for *Phragmites*, *Typha* and papyrus. *Cyperus immensus* was frequently attacked by vermin monkeys and only attained a maximum biomass of 146 g/m^2 by July 2003. It was thus removed from the cells, and, thereafter the cells were left as unplanted controls.

The relative growth rates (RGR day^{-1}) for the remaining macrophyte species at different stages of wetland operation are presented in Figure 4.3. *Phragmites* maintained a positive RGR throughout the experiment. *C. papyrus* growth declined in the period May 2003 up to July 2004 (partly 3-day batch operation) but picked up again during the period July – December 2004 (continuous flow phase 1). Both *Phragmites* and *Typha* showed signs of recovery (increase in RGR) with the 3-day batch operation (cf. Figure 4.1 above). There was a marked decline in RGR during the 5-day batch operation in the plant shoot re-growth period. Many papyrus umbels dried up due to desiccation in the same period.

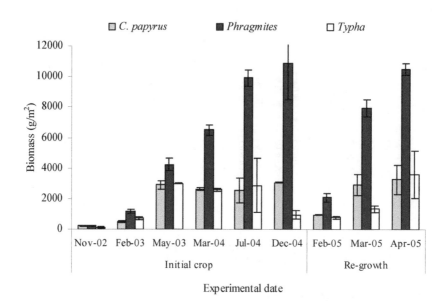

Fig. 4.2 Aerial biomass production (as dry weight) during the initial crop and in the re-growth stage. In December 2004 all above ground biomass was harvested.

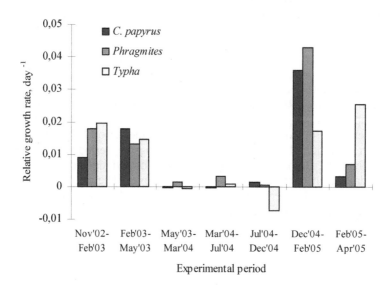

Fig. 4.3 Plant relative growth rate under various wetland operation modes. November 2002-July 2003 = Batch operation 5-day; July 2003 to March 2004 = Batch 3-day; March 2004-February 2005 = continuous flow phase 1 and February-April 2005 = Batch 5-day operation. *Typha* cells were loaded on continuous flow from 4th March 2005.

Below ground biomass
At the start of the study average below ground biomass based on dry weight per unit area were 57 ± 19 g/m^2, 187 ± 125 g/m^2 and 389 ± 44 g/m^2 (n=6) for *Typha*, *Phragmites* and *C. papyrus* respectively. By the end of the study (after 33 months) the overall below ground biomass yields were 1834 g/m^2, 2374 g/m^2 and 2045 g/m^2 respectively for the same macrophyte species. *Typha*'s root and rhizomes contributed 31 % of the total biomass in the cells both at the beginning and at the end of the experiment. However, for *Phragmites* and papyrus there was a decrease in the proportion of rooting biomass (root/shoot ratio) from 44 % and 63 % respectively at the start to 37 % and 46 % respectively at the end of the study. Macrophyte root penetration into the substratum was shallow in the case of papyrus with a maximum depth of 10 cm while that of *Phragmites* was 20 cm. In contrast, *Typha* roots penetrated right to the bottom (30 cm) in both cells forming a fine mesh enclosing the substratum.

Nutrient storage and uptake rates

Above ground plant tissue nitrogen and phosphorus were variable in the different stages of growth and wetland operation (Figures 4.4 and 4.5). Overall *Phragmites* shoots had the highest (9.2 ± 0.7 mg/g) nitrogen concentration compared to *Typha* (7.4 ± 0.5 mg/g) and papyrus (6.1 ± 0.2 mg/g). On the other hand *Typha* had higher phosphorus concentration (1.9 ± 0.11mg/g) when compared to *Phragmites* (1.7 ± 0.12 mg/g) and papyrus (1.6 ± 0.14 mg/g). Nitrogen to phosphorus ratio in shoots was highest in *Phragmites* (5.4:1). Of all plant parts, juvenile shoots and inflorescence had the highest nutrient concentrations (Table 4.4). Senescing plant shoots (about 10 - 50% achlorophyllous (chlorotic) had significantly lower nutrient concentrations. For instance, senescing papyrus culms and umbels had mean phosphorus concentrations of 0.52 ± 0.013 mg/g and 1.1 ± 0.006 mg/g respectively while nitrogen concentrations were 4.4 ± 0.45 mg/g and 4.2 ± 0.39 mg/g respectively. Senescing *Typha* shoots on the other hand had 0.89 ± 0.14 mg/g phosphorus and 2.71 ± 1.01 mg/g nitrogen. Rhizomes and roots had lower nitrogen and phosphorus concentration after three months (February 2003) than at the start of the study (November 2002). Nitrogen concentrations in the rhizome and roots of *Phragmites* and *Typha* for instance, declined by about 50 %.

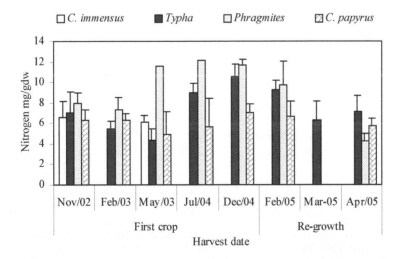

Fig. 4.4 Nitrogen concentrations (based on dry weight) in plant shoots at various stages of growth and wetland operation (n = 6 - 9).

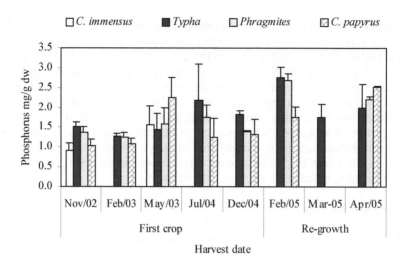

Fig. 4.5 Phosphorus concentrations (based on dry weight) in plant shoots at various stages of growth and wetland operation (n = 6 - 9).

Table 4.4 Nutrient concentrations in juvenile shoots and in various mature plant organs (based on dry weight). Superscripts 1- 4 refer to natural wetlands from which young shoots or cuttings were obtained for planting in the constructed wetland.

Plant species	Plant age	Plant part	Study wetland		Natural wetland	
			P mg/g	N mg/g	P mg/g	N mg/g
[1]*Cyperus immensus*	juvenile	shoot	2.4 - 3.8	7.7		
	mature	shoot	1.6 - 1.8	6.1 - 6.2	0.9	6.6
		inflorescence	2.8 - 3.9	8.8		
		Stolon+root	0.7 - 1.5	2.3		
[2]*Cyperus papyrus*	juvenile	shoot	1.0 – 1.5	5.3 – 8.9	1.8	6.1
	mature	culm	1.0 – 2.6	3.9 – 5.7	2.0	4.1
		umbel	1.8- 2.5	8.5 – 11.1	2.1	10.7
		Rhizome+root	0.9	4.5		
[3]*Typha domingensis*	juvenile	shoot	1.1 – 3.6	5.8 – 11.7	2.1	8.3
	mature	shoot	1.5 – 2.8	4.4 – 10.5	2.6[*]	12.9[*]
		inflorescence	5.8	11.8		
		Rhizome+root	1.6 - 1.7	3.2 - 6.0		
[4]*Phragmites mauritianus*	juvenile	shoot	1.2 – 3.2	6.2 – 10.6	1.5	14
	mature	shoot	1.4 – 2.7	5.8 – 11.7	2.1	9.4
		Rhizome+root	1.1 - 1.7	3.7 – 7.8		

The natural wetlands were frequently cultivated for food crops or cattle-grazed. Sources: [1] = Lake Kanyaboli swamp (Siaya district); [2] = Auji-Dunga swamp (Kisumu bay), [3] = K'Achok swamp (Kisumu), * denotes *Typha* plants previously growing on part of the first aeration pond of PANPAPER Mills, [4] = Mudurme river swamp (Siaya district).

Aerial nutrient storage was variable over the experimental period (Table 4.5). Nutrient accumulation rates were higher during periods of high biomass production at the exponential growth stage in the initial crop (February to May 2003) and during re-growth in March to April 2005 (Figure 4.2). Nutrient storage in shoots was similarly high when plant tissue nutrient concentration was high (cf Figures 4.4 - 4.5). The maximum nitrogen uptake rates for *Typha*, *Phragmites* and papyrus were 0.102 g/m^{2}day, 0.365 g/m^{2}day, and 0.121 g/m^{2}day respectively. Peak phosphorus uptake rates were 2 to 7 times lower.

Table 4.5 Phosphorus and nitrogen storage (g/m^2) and uptake rates (g/m^2.day) in plant shoots in various periods

	Phosphorus					
Period	*Typha*		*Phragmites*		papyrus	
	g/m^2	g/m^2.day	g/m^2	g/m^2.day	g/m^2	g/m^2.day
Nov'02 - Feb'03	0,92	0,009	1,47	0,013	0,54	0,003
Feb'03 - May'03	4,37	0,034	6,73	0,050	6,51	0,055
Mar'04 - Jul'04	6,25	0,005	17,28	0,045	3,14	-0,001
Jul'04 - Dec'04	1,71	-0,023	15,04	0,009	4,03	0,005
Dec'04 - Feb'05	2,17		5,65		1,66	
Feb'05 - Mar'05	2,36	0,031				
Mar'05 - Apr'05	7,19	0,103	23,07	0,126	8,26	0,021
	Nitrogen					
	Typha		*Phragmites*		papyrus	
	g/m2	g/m^2.day	g/m^2	g/m^2.day	g/m^2	g/m^2.day
Nov'02 - Feb'03	4,01	0,037	8,63	0,077	3,13	0,019
Feb'03 - May'03	13,12	0,102	49,39	0,365	14,31	0,121
Mar'04 - Jul'04	25,84	0,019	120,14	0,312	14,44	-0,004
Jul'04 - Dec'04	9,95	-0,136	127,27	0,075	21,80	0,026
Dec'04 - Feb'05	7,28		20,41		6,31	
Feb'05 - Mar'05	8,44	0,112				
Mar'05 - Apr'05	25,52	0,364	45,09	0,246	18,82	0,048

Nutrient removal efficiency

Wetland inflow wastewater quality was variable but on average similar with respect to nitrogen and phosphorus concentrations (Table 4.6). Short-term increases in both nitrogen and phosphorus were experienced during the de-sludging of PANPAPER Mills secondary treatment ponds. Mean total nitrogen in the wetland inflow was in the range 2.1 – 2.9 mg/l with total dissolved nitrogen making up 49 – 57 %. Total phosphorus was in the range 0.54 – 0.79 mg/l with the total dissolved component being only 18 – 30%.

Nitrogen and phosphorus concentrations in the wetland outflow were highly variable throughout the study with occasional net release. Figures 4.6 and 4.7 depict the variation in continuous flow phases 1 and 2 respectively. Mass removal efficiency for nitrogen was higher than that of phosphorus throughout the study period. Overall removal efficiencies were higher when the wetland system had continuous flow than during batch loading.

Table 4.6 Characteristics of the wetland inflow wastewater in batch (phases 1 – 3) and continuous flow (phase1&2) operation modes. Loading rates are in g/m^3.day

Parameter		Batch			Continuous flow	
		Phase 1	Phase 2	Phase 3	Phase 1	Phase 2
TP	range	0.56-0.91	0.36-0.83	0.46-0.77	0.4-1.1	0.36-0.72
	Mean/se	0.72±0.01	0.67±0.02	0.59±0.02	0.79±0.02	0.54±0.03
	Loading rate	0,06	0,09	0,05	0,13	0,11
	n	32	31	15	36	13
TDP	range	0.02-0.22	0.10-0.30	0.05-0.35	0.07-0.71	0.08-0.27
	Mean/se	0.13±0.02	0.16±0.02	0.14±0.02	0.20±0.06	0.16±0.02
	Loading rate	0,01	0,02	0,01	0,03	0,03
	n	10	10	15	11	11
TN	range	1.7-4.0	1.67-4.4	1.01-3.5	0.46-5.5	1.55-3.1
	Mean/se	2.9±0.11	2.9±0.13	2.45±0.15	2.1±0.27	2.4±0.11
	Loading rate	0,23	0,41	0,20	0,36	0,48
	n	35	32	15	29	13
TDN	range	0.99-2.42	1.07-2.41	0.69-2.25	0.23-2.6	0.61-1.9
	Mean/se	1.6±0.13	1.67±0.09	1.2±0.095	1.2±0.15	1.27±0.11
	Loading rate	0,13	0,23	0,10	0,20	0,25
	n	10	20	15	22	11

Continuous flow

Total nitrogen removal efficiency was highest during the exponential growth stage (continuous flow phase 2). *Typha* cells had 74.9±2.6 % mean removal efficiency while *Phragmites* cells had a mean of 70.8±2.2 %. Papyrus cells achieved 51.2±4.4 % while the controls had 43.7±3.3 % removal efficiency. The higher loaded series E *Typha* cells had 57.4±4.3 % removal, nearly the same as that from papyrus cells. In the steady growth stage (continuous flow phase 1) the performance of the control cells remained nearly the same with 42.4 % removal efficiency while that of the planted wetland cells was 68±3.1 %, 61.9±4.2 % and 49.3±5.1 % for *Typha*, *Phragmites* and papyrus respectively.

Total phosphorus removal efficiency was more variable and lower than that of nitrogen. During the exponential growth stage (continuous flow phase 2) *Typha* cells had significantly higher removal efficiency (61.5±4.4 %) compared to *Phragmites* (45.6±2.1 %) and papyrus (29.6±6.8 %). The control cells had a net release of phosphorus (-4.3±11 %). The higher loaded series E *Typha* cells achieved 27.6±5.0 % removal. In the steady growth stage the control cells had a removal efficiency of 38.3±2.9 %. There was no significant difference ($p < 0.05$) in the performance of both *Typha* and *Phragmites*. Their mean removal efficiencies were 49.1±2.5 % and 50.5±3.2 % respectively. Papyrus had the lowest phosphorus removal (36.5±2.5 %).

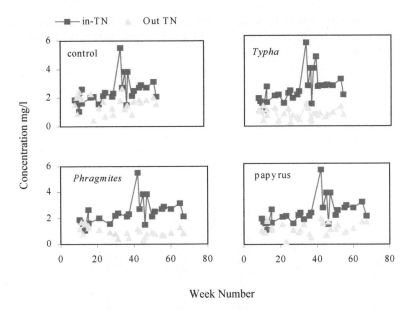

Fig. 4.6 Temporal variations in inflow and outflow total nitrogen in various wetland cells in continuous flow phase 1 (week no 10 - 47) and phase 2 (week no. 48 - 67). Week no. 10 = 16 - 22 May 2004. The control cells contain substrate without plants.

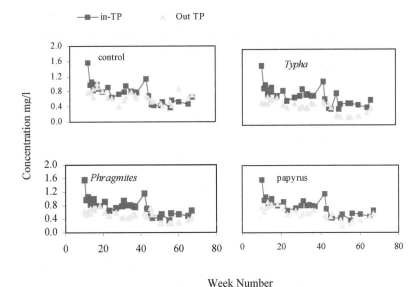

Fig. 4.7 Temporal variations in inflow and outflow total phosphorus in various wetland cells in continuous flow phase 1 (week no 10 - 47) and phase 2 (week no. 48 - 67). Week no. 10 = 16 - 22 May 2004. The control cells contain substrate without plants.

Batch operation

As in continuous flow nitrogen removal efficiency was higher than phosphorus removal (Figures 4.8 and 4.9). Nitrogen removal was significantly higher (p<0.05) in *Typha* and *Phragmites* than in papyrus cells at the longer retention time (phases 1 and 3).

In phase 2 (3-day retention time) however, there was no significant difference in removal efficiency of cells with different plant species. Phragmites cells maintained constant removal efficiency in all phases.

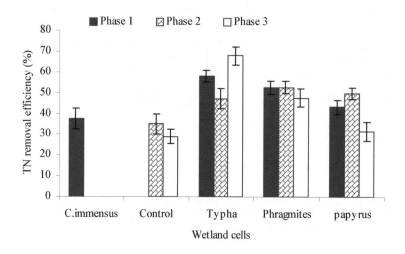

Fig. 4.8 Total nitrogen removal efficiencies in various wetland cells during batch loading. Phases 1 - 3 were operated in the periods November 2002 - July 2003, July 2003 - March 004 and February - April 2005 respectively.

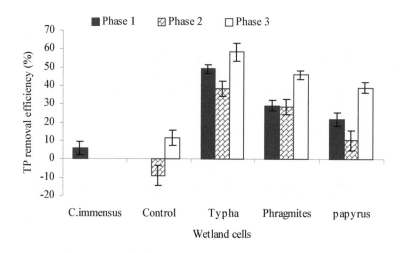

Fig. 4.9 Total phosphorus removal efficiencies in various wetland cells during batch loading. Phases 1 - 3 were operated in the periods November 2002 - July 2003, July 2003 - March 2004 and February - April 2005 respectively.

The control cells had the lowest nitrogen removal efficiency in all phases. However, in phase 3 their performance was not significantly different from that of papyrus cells. Outflow concentrations were in the range of 0.97 – 1.7 mg N /l (Table 4.7).

Total phosphorus removal efficiency was significantly higher for all cells at the longer retention time. *Typha* cells had the highest removal efficiency in all phases while *C. immensus* cells (phase 1) had the lowest among the planted cells. The highest removal in papyrus cells occurred in phase 3. The control cells had low removal efficiency giving a net release of phosphorus in phase 2. Wetland outflow concentrations were in the range 0.27 – 0.66 mg/l (Table 4.7).

Table 4.7 Total nitrogen and phosphorus concentrations in wetland outflow during batch operation (n = 26 – 32).

Parameter	Cell type	Phase 1 5-day	Phase 2 3-day	Phase 3 5-day
TN mg/l	*C. immensus*[1]	1.33 ±0.075	-	-
	Typha[2]	0.97±0.12	1.2±0.1	0.99±0.12
	Phragmites	1.10±0.07	1.1 ±0.07	1.1±0.06
	C. papyrus	1.32±0.08	1.2±0.08	1.5±0.08
	Control	-	1.45±0.08	1.7±0.11
TP mg/l	*C. immensus*[1]	0.55±0.017	-	-
	Typha[2]	0.29±0.013	0.39±0.016	0.27±0.027
	Phragmites	0.40±0.018	0.45±0.014	0.28±0.008
	C. papyrus	0.44±0.015	0.57±0.018	0.31±0.008
	Control	-	0.66±0.027	0.46±0.017

[1] n = 16 for *C. immensus* (phase 1); [2] n = 8 for *Typha* (phase 3)

Nutrient mass flow
During the exponential growth stage of the initial crops both phosphorus and nitrogen uptake by plant shoots in all wetland cells exceeded inputs by inflow wastewater in the period under consideration except for nitrogen in *Typha* shoots (Table 4.8). In the latter case nitrogen uptake was 92 % of input. During the steady growth stage phosphorus uptake by *Typha* and *Phragmites* shoots were 9 % and 79 % of the inputs from wastewater respectively. Papyrus shoots had a net release of 3 mg P/day. Nitrogen uptake by *Typha* shoots was 13.3% of the input by inflow wastewater. *Phragmites* nitrogen uptake exceeded the input by inflow wastewater while papyrus had a net release of 10 mg N/day.

Table 4.8 Phosphorus and nitrogen mass flows (g/day) for the steady growth (March-July 2004) and exponential growth stages (February – May 2003). The periods correspond to months 3 - 6 and 16 - 20 of operation respectively. The contributions of sediments and litter accumulated prior to the periods are excluded.

	Phosphorus			Nitrogen		
	Typha	*Phragmites*	papyrus	*Typha*	*Phragmites*	papyrus
	Exponential growth stage					
Inflow	0,068	0,068	0,061	0,276	0,276	0,247
Outflow	0,032	0,047	0,048	0,285	0,447	0,578
Shoot uptake	**0,085**	**0,125**	**0,138**	0,255	**0,913**	**0,303**
	Steady growth stage					
Inflow	0,134	0,142	0,142	0,357	0,378	0,378
Outflow	0,032	0,047	0,048	0,102	0,13	0,168
Shoot uptake	0,013	0,113	-0,003	0,048	**0,78**	-0,01

Bold face amounts indicate that plant shoot uptake exceeded input by inflow wastewater in the periods considered.

Discussion

Plant vitality, biomass production and nutrient uptake

Plant shoots appeared greener during continuous flow than at batch loading. At the end of each batch cycle wetland cells were drained for up to three hours before reloading with fresh wastewater. This may have stressed the plants especially *Cyperus papyrus*. Whereas papyrus shoots were greener in the re-growth period, desiccation of umbels and sometimes whole shoots occurred under batch operation.

The maximum shoot heights attained by *Phragmites*, *Typha* and papyrus plants in this study, 4.9 m, 4.4 m and 3.4 m respectively were within the ranges for similar plants in the natural wetlands from which young shoots or cuttings had been collected for planting. However, for papyrus the shoot height was on the lower end of the range (3 - 5 m) for the source natural wetland (Auji-Dunga swamp in Kisumu bay). Similar low shoot heights were reported for "stunted" papyrus in the western edge of the Nakivubo swamp in Uganda (Kansiime and Nalubega, 1999).

The maximum aerial standing biomass of the initial mature crop of *Phragmites mauritianus* (10896 g/m^2) and *Typha domingensis* (3015 g/m^2) were higher (200 -300 % respectively) than the maximum reported by Balirwa (1998) for wetlands fringing Lake Victoria at Napoleon Gulf, Uganda and by Thompson (1985). *Typha*'s biomass was lower than that reported for *Typha latifolia* (4400 g/m^2) by Thut (1993) in a similar size constructed wetland receiving secondary-treated pulp mill effluent in Mississippi, U.S.A. Although the wetland had similar nutrient concentrations (3 mg N/l, 0.6 mg P/l) in the inflow wastewater to those in our study it had nearly double our planting density and 3 - 5 times higher hydraulic loading rates. The biomass for papyrus (3075 g/m^2) was 30 - 50 % lower than those reported for various natural wetlands by Denny (1984), Jones and Muthuri (1997) and Balirwa (1998) but slightly higher than that (1983 - 2862 g/m^2) reported by Kipkemboi *et al.* (2002) for natural wetlands on the northern shores of Lake Victoria in Uganda.

The standing biomass productivity of the wetland cells was variable. The overall above ground biomass productivity computed for the period up to the steady growth stage was highest in *Phragmites* cells (210 kg/ha.day). This is close to that obtained for *Phragmites mauritianus* (256 kg/ha.day) in a constructed wetland receiving pre-settled municipal wastewater in Jinja, Uganda (Okurut, 2001) and *Phragmites australis* (192 kg/ha.day) in a subsurface flow wetland receiving sewage in Groningen, The Netherlands (Meuleman, 1999). The productivity of papyrus (143 kg/ha.day) obtained in this study is lower than those found in some natural wetlands. For example, Jones and Muthuri (1997) reported a net productivity of 6.28 kg/m^2.day (172 kg/ha.day) in a swamp located on Lake Naivasha, Kenya while Kansiime *et al* (2003) reported a productivity of 210 kg/ha.day in areas that were not under the influence of wastewater in the Nakivubo swamp, Uganda. However, the data reported by Kansiime *et al* (2003) was taken in a five-month period of papyrus re-growth. The re-growth rate for papyrus in our study was comparable (250 kg/ha.day). *Typha*'s productivity of 154 kg/ha.day (ca 56 t/ha.yr) found in this study is double the highest total net productivity (25 – 30 t/ha.yr) suggested by Howard-Williams and Gaudet (1985).

The productivity to shoot biomass ratios were 1.4, 1.9 and 1.7 respectively for *Phragmites*, *Typha* and papyrus indicating that *Phragmites* suffered less losses than the other macrophytes. This is further supported by positive relative growth rates in the latter maintained throughout the study. *Typha* plants matured earlier than the rest and started senescing by July 2004 while *Phragmites* and papyrus continued to increase in biomass. Papyrus however was affected by periods of dryness in the load-drain batch operation mode, recovering at the commencement of continuous flow (Figures 4.1 - 4.3).

Plant tissue nutrient concentrations and nutrient standing stock varied between species, growth stages and mode of wetland operation. Nutrient concentrations found for plant shoots especially under continuous flow were similar to those of their source natural wetlands (Table 4.4) but lower than those reported for other natural wetlands. For example, Muthuri and Jones (1997) found concentrations of 11.6 mg N/g and 1.6 mg P/g in mature whole papyrus shoots in a swamp fringing Lake Naivasha, Kenya. Their phosphorous concentration is the same as that found in our study but their nitrogen concentration is nearly double. The reason for this discrepancy is two-fold. On the one hand, the young shoots (propagules) that were planted in the study wetland, for practical reasons were dug out from wetland edges where there was no standing water. Mature shoots were simultaneously harvested for comparative nutrient determination. Such wetland edges may have stunted plants (Kansiime *et al.*, 2003) with low nutrient concentrations. On the other hand, nutrient concentrations in the inflow wastewater were low. In the initial crop, the study wetland had not yet accumulated a high enough nutrient pool. It is expected that this would increase with wetland age. Tissue nutrient concentrations were not determined after April 2005. However, from the higher shoot density recorded in July 2005 it may be inferred that biomass increased above the maximum previously recorded (April 2005) and that aerial nutrient uptake rates and tissue concentrations would have been higher.

Nutrient mass flows

The mass flows for nitrogen and phosphorus (Table 4.8) indicate that plant uptake plays a significant role in removing nutrients from the wastewater of a low loaded system such as the one in this study. Toet *et al.* (2005) made similar conclusions for a treatment wetland used for polishing treated sewage in The Netherlands. However, the nutrient amount removed depends on the growth stage of the plants. Plant uptake exceeded inputs by inflow wastewater in the exponential growth stage. Considering that dissolved nutrient ions in the inflow wastewater were not more than 30 % and 57 % of the totals for phosphorus and nitrogen respectively (Table 4.6) it is evident that plants took up additional nutrients from different compartments of the wetland. These include the sediment pool (decomposing organic matter) accumulated prior to the period under consideration, relocation from senescing plant parts and decomposing litter. Denny (1987) alluded to close cycling of mineral ions in wetlands with limiting external nutrient input. Additional nitrogen may be supplied by fixation in the rhizosphere (Gaudet,

1979). In the steady growth stage *Phragmites* shoots had higher nutrient uptake than *Typha* while papyrus had a net release. The differences are largely due to the phenology of the plants.

Since the growth (biomass production) of *Typha* and *Phragmites* plants were comparable to or exceeded those reported for other natural wetlands, the shoot uptake (g/day) at exponential growth stage may be considered to represent the minimum flows through the wetland necessary for healthy plant growth and optimal wetland performance. For papyrus however, higher budgets are necessary. These may be achieved by increasing the hydraulic loading rate and/or the effective wetland volume. The latter may be achieved by omitting gravel since papyrus can float.

Nutrient removal efficiency and implications for water quality of the Nzoia River

Mean removal efficiencies for total nitrogen were in the range of 49 - 75 % for planted cells and 42 - 49% for unplanted ones in continuous flow. Series E *Typha* cells with nearly double the hydraulic loading rate of other cells achieved 57 % removal. During batch operation modes nitrogen removal efficiencies were in the range of 32 - 68 % for planted cells (excluding *C. immensus*) and 29 - 35 % for the unplanted. The same trend was observed for phosphorus removal with efficiencies of 30 - 60 % in planted cells and -4 – 38 % in unplanted ones in continuous flow compared to 10 – 58 % in planted and -9 - 12 % in unplanted cells under batch operation. The lower removal efficiencies under the batch load-drain operation mode may be attributed to re-suspension of previously deposited solids in the gravel voids (Kadlec and Knight, 1996). Solids re-suspension in the wetland also resulted in higher total suspended solids in the outflow wastewater (Chapter 5).

Planted cells achieved higher removal of both nitrogen and phosphorus from the inflow wastewater compared to unplanted ones. From their differences, up to 25 % removal efficiency may be attributed to the presence of plants. Besides direct uptake, plant roots and rhizomes provide extra surface area for attachment of microorganisms and therefore enhance nutrient removal via microbial-mediated processes. The rooting structure enhances removal of particulates that are then retained in the wetland bed. The influence of plants and their growth on wetland performance with regard to other pollutants is discussed in Chapters 5 and 6.

The concentrations of nitrogen and phosphorus discharged in the treated effluent by PANPAPER Mills are low (Chapter 2). However, due to the large quantity of effluent discharged there is a potential for enrichment especially in impoundments downstream, including Lake Victoria. As discussed in Chapter 2 current discharge from PANPAPER Mills increases the concentration of nitrogen downstream by 50 – 100 %. If a constructed wetland planted with *Typha domingensis* or *Phragmites mauritianus* and having the same performance as in this study (removal 62 - 75 % TN and 49 - 62 % TP) were integrated with the present PANPAPER Mills treatment ponds we predict that at low flow in the receiving River Nzoia, nitrogen concentration downstream of discharge would be 0.87 mg/l remaining nearly the same as that in the upstream location (0.84 mg/l, Chapter 2) indicating no net pollution. Current predicted phosphorus concentrations in the river downstream are 20 % higher than in the upstream site. With the wetland integration the predicted concentration downstream would be 10 % higher than upstream implying a reduction in pollution by 50 % of the current level in the river.

Conclusion

From the findings of this study and the foregoing discussion we conclude as follows: In our constructed wetland pilot project for the treatment of wastewater from a pulp and paper mill in Kenya,

1. Plant tissue nutrient concentrations were lower than in healthy natural wetland plants. Nevertheless, *Typha* and *Phragmites* had satisfactory biomass production. The growth of papyrus was sub-optimal.
2. *Phragmites, Typha* and papyrus (in that order) are all suitable species for the effective removal of nutrients from pulp and paper mill wastewater in East Africa. It is suggested that a mixture of all three species (or whichever is available in the district) might provide the best long-term option.
3. Plant vitality and growth and wetland performance with respect to nutrient removal were lower during the batch mode of wetland operation.
4. Nutrient mass flows indicate that in a low loaded system such as the one in this study, cycling of nutrients in sediments and/or in senescing/decaying plant organs are an important source for sustaining plant growth. Therefore harvesting of shoots should be appropriately timed to avoid depletion of the nutrient pool.
5. Integrating a constructed wetland as a tertiary treatment stage with the PANPAPER Mills wastewater ponds would eliminate pollution of river Nzoia with respect to nitrogen concentration and reduce pollution with phosphorus by 50 % of the current level in the river.

References

Abira, M. A., Ngirigacha, H. W. and van Bruggen J.J.A. 2003. Preliminary investigation of the potential of four tropical emergent macrophytes for treatment of pre-treated pulp and papermill wastewater in Kenya. *Water Science and Technology* 48 (5): 223 - 231

APHA, 1995. *Standard methods for the analysis of water and wastewater*, 19th edition. American Public Health Association, Washington DC.

Balirwa, J.S. 1998. Lake Victoria wetlands and the ecology of the Nile Tilapia, *Oreochromis niloticus* Linné. Ph.D. Thesis, Balkema publishers, Rotterdam, The Netherlands, 245 pp.

Bett, A.C. 2006. Assessment of nutrient removal by constructed wetlands in Lake Naivasha, Kenya. M.Sc. Thesis ES 06.11, UNESCO-IHE, Delft, The Netherlands.

Brix, H. 1993. Macrophyte-mediated oxygen transfer in wetlands: Transport mechanisms and rates. In: Moshiri, G.A. (ed.). *Constructed wetlands for water quality improvement*. Lewis publishers, Boca Raton, pp 391 - 398.

Brix, H. 1994. Functions of macrophytes in constructed wetlands. *Water Science and Technology* 29 (4): 71 – 78.

Brix, H., 1997. Do macrophytes play a role in constructed treatment wetlands? *Water Science and Technology* 35 (5): 11 - 17.

Denny, P. 1984. Permanent swamp vegetation of the Upper Nile. *Hydrobiologia* 110: 79 - 90.

Denny, P. 1987. Mineral cycling by plants: a review. *Archiv f. Hydrobiologie Beih.* 27: 1-25.

Denny, P. 1993. Wetlands of Africa. In: Whigham, D.F., Dykyjova and Hejny, S (Eds.) *Wetlands of the world 1. Inventory, ecology and management.* Kluwer Academic Publishers, Dodrecht, pp 1-128.

Gaudet, J.J. 1979. Seasonal changes in nutrients in a tropical swamp: North Swamp, Lake Naivasha, Kenya. *Journal of Ecology* 67: 953 - 981.

Greenway, M. 1997. Nutrient content of wetland plants in constructed wetlands receiving municipal effluent in tropical Australia. *Water Science and Technology* 35 (5): 15 – 142.

Gunstenspergen, G.R., Stearns, F. and Kadlec, J.A., 1989. Wetland vegetation. In: Hammer D. A. (ed.), *Constructed wetlands for wastewater treatment: municipal, industrial and agricultural.* Lewis Publishers, Chelsea, Michigan, U.S.A., pp. 73 - 81.

Howard-Williams, C and Gaudet, J.J. 1985. The structure and functioning of African swamps. In: Denny, P. (Ed.): *The ecology and management of African wetland vegetation*. Dr.W. Junk publishers, Dordrecht, pp. 153 - 175.

JICA/GOK, 1992. *National Water Master Plan, Data Book: Hydrological data*. Japan International Cooperation Agency and Government of Kenya, Ministry of Water Development.

Jones, M.B. and Muthuri, F.M. 1997. Standing biomass and carbon distribution in a papyrus (*Cyperus papyrus* L.) swamp on Lake Naivasha, Kenya. *Journal of Tropical Ecology* 13: 347 – 356.

Juwarkar, A.S., Oke, B., Juwakar, A. and Patnaik, S.M. 1995. Domestic wastewater treatment through constructed wetland in India. *Water Science and Technology* 32 (3): 291 - 294.

Kadlec, R.H. and Knight, R.L. 1996. *Treatment wetlands*. CRC Press Inc., Lewis Publishers, Boca Raton, Florida, 893 pp.

Kadlec, R.H., Knight, R.L., Vymazal, J., Brix, H., Cooper, P. and Haberl, R., 2000. *Constructed wetlands for pollution control: processes, performance, design, and operation*. IWA specialist group on use of macrophytes in water pollution control. Scientific and technical report No.8, IWA publishing, London, UK, 156 pp.

Kansiime, F. and Nalubega, M. 1999. Wastewater treatment by a natural wetland: the Nakivubo swamp, Uganda. Processes and implications. Ph.D. thesis, A.A. Balkema Publishers, Rotterdam, The Netherlands, 300 pp.

Kansiime, F., Nalubega, M., van Bruggen, J.J.A. and Denny, P. 2003. The effect of wastewater discharge on biomass production and nutrient content of *Cyperus papyrus* and *Miscanthidium violaceum* in the Nakivubo wetland, Kampala, Uganda. *Water Science and Technology* 48 (5): 233 - 240.

Kipkemboi, J., Kansiime, F. and Denny, P. 2002. The response of *Cyperus papyrus* (L.) and *Miscanthidium violceum* (K. Schum.) Robyns to eutrophication in natural wetlands of Lake Victoria, Uganda. *African Journal of Aquatic Sciences*, 27: 11 - 20.

Koottatep, T. and Polprasert, C. 1997. Role of plant uptake on nitrogen removal in constructed wetlands located in the tropics. *Water Science and Technology* 37 (12): 1 - 8.

Kvet J. and Westlake D.F. 1983. Primary Production in Wetlands. In: *The Production Ecology of Wetlands*, Westlake D.F., Kvet J. and Szczepanski, A. (eds.) Cambridge University Press, pp. 78 - 168.

Muthuri, F.M. and Jones, M.B. 1997. Nutrient distribution in a papyrus swamp: Lake Naivasha, Kenya. *Aquatic Botany* 56: 35- 50.

Meuleman, A.F.M., 1999. Performance of treatment wetlands. Ph.D. thesis, Utrecht University, The Netherlands.

Novozamsky I., Houba V.G., van Eck R. and van Verk, W. (1983). A Novel Digestion Technique for Multi-element Plant Analysis. *Communication in Soil Science* 14: 239 – 248.

Nyakang'o, J.B. and van Bruggen, J.J.A. 1999. Combination of a well functioning constructed wetland with a pleasing landscape design in Nairobi, Kenya. *Water Science and Technology* 40 (3): 249 - 256.

Okurut, T. O. 2000. A pilot study on municipal wastewater treatment using a constructed wetland in Uganda. PhD dissertation, Balkema publishers, Rotterdam, The Netherlands

Okurut, T. O. 2001. Plant growth and nutrient uptake in a tropical constructed wetland. In: Vymazal, J. (Ed.). *Transformations of nutrients in natural and constructed wetlands*. Buckhuys Publishers, Leiden, The Netherlands, pp 451 - 462.

Reddy, K.R., Patrick Jr., W.H. and Lindau, C.W. 1989. Nitrification-denitrification at the plant-root sediment interface in wetlands. *Limnology Oceanography* 34: 1004 - 1013.

Senzia, M.A.; Mashauri, D.A. and Mayo, A.W. 2002. Suitability of constructed wetlands and waste stabilisation pond in wastewater treatment: nitrogen transformation and removal. 3rd WaterNet/Warfsa symposium on water demand management for sustainable development, Dar es Salaam, October 2002. http://www.waternet.ihe.nl/aboutWN/pdf/Senzia&al.pdf [2003, Jul 24]

Tanner, C. C. 1996. Plants for constructed wetland treatment systems-A comparison of the growth and nutrient uptake of eight emergent species. *Ecological Engineering* 7: 59 – 83.

Thompson, K., 1985. Emergent plants of permanent and seasonally flooded wetlands. In: Denny, P. (ed.) *The ecology and management of African wetland vegetation*. Junk Publishers, Dordrecht, The Netherlands, pp. 43 - 107.

Toet, S., Bouwman, M., Cevaal, A. and Verhoeven, J.T.A. 2005. Nutrient removal through autumn harvest of *Phragmites australis* and *Typha latifolia* shoots in relation to nutrient loading in a wetland system used for polishing sewage treatment plant effluent. *Journal of Environmental Science and Health part A* 40 (6-7): 1133 - 1156.

Thut, R.N., 1989. Utilisation of artificial marshes for the treatment of pulp mill effluents. In: Hammer, D.A. (ed), *Constructed wetlands for wastewater treatment: municipal, industrial, and agricultural*. Lewis, Chelsea, Michigan, U.S.A., pp 239 –244.

Thut, R.N., 1993. Feasibility of treating pulp mill effluent with a constructed wetland. In: Moshiri, G.A. (ed.). *Constructed wetlands for water quality improvement*, CRC Lewis Publishers, Boca Raton, , pp 441 - 447.

Vymazal, J. 2001. Types of constructed wetlands for wastewater treatment: their potential for nutrient removal. In: Vymazal, J. (ed.). *Transformations of nutrients in natural and constructed wetlands*. Buckhuys Publishers, Leiden, The Netherlands, pp 1 – 93.

Chapter five

The efficacy of the constructed wetland in removing organic matter and suspended solids

The efficacy of the constructed wetland in removing organic matter and suspended solids

Abstract

The use of constructed wetlands to improve water quality has spread to many fields including industrial effluent treatment. A pilot-scale study was undertaken at the Pan African Paper Mills (E.A.) Ltd. in the western highlands of Kenya to establish the efficiency of a subsurface flow constructed wetland in purifying pre-treated pulp and paper mill wastewater with respect to organic matter (BOD, COD) and suspended solids (TSS). The study further determined the reaction rate parameters and residuals for BOD removal. The wetland operation involved varying hydraulic loading rates, retention times, operation modes and macrophyte growth conditions.

The constructed wetland effectively removed BOD (up to 90 %) and TSS (up to 81 %) from the wastewater to concentrations below the national guidelines in Kenya. However, COD removal was low (up to 52 %) and the guideline of 100 mg/l for it was not achieved in this study. The wetland's operation mode influenced COD and TSS removal but not the removal of BOD. COD and TSS removal efficiency were significantly higher in continuous flow. In batch operation mode, BOD removal was significantly higher at the longer retention time. For COD and TSS however, there was no significant difference at the two retention times. The non-zero background concentration for BOD varied between 4.3 and 7.4 mg/l for the different cells while areal BOD reaction rate constants varied from 0.055 – 0.114 m/day (20 – 42 m/yr). The reaction rates are reported for pulp and paper mill wastewater for the first time.

Presence or absence of plants did not influence BOD removal. However, COD removal efficiency was significantly higher in unplanted compared to planted cells when plants were in a steady growth stage. For TSS, although removal efficiencies were higher when plants were in exponential growth stage, there was no significant difference between planted and unplanted cells except for *Typha* cells which had better performance than the controls. *Typha* cells had consistently higher TSS removal efficiency than *Phragmites* and papyrus in continuous flow. This may be due to differences in plant root penetration depth. *Typha* roots penetrated the entire bed depth and therefore afforded a better filtration medium. It may be noted that besides TSS removal in the wetland beds with developed plant roots, the presence of macrophytes does not enhance BOD and COD removal when compared to unplanted cells.

Key words

Efficiency, BOD, COD, TSS, reaction rate, pulp and paper mill wastewater

Introduction

Water pollution problems associated with industries have become a common feature of the Lake Victoria catchment area in the western part of Kenya. One of the industries, the Pan African Paper Mills (E.A.) Limited (PANPAPER) with an estimated average discharge of 35,000 m^3 per day, contributes the lion's share of all effluents discharged into the Nzoia river. The Mills' effluent is clarified and treated in lagoons before discharge. However, the final effluent quality seldom complies with government-prescribed effluent discharge guidelines for organic matter and suspended solids. Discharge of the effluent increases organic matter (as COD – chemical oxygen demand) by up to 100 % while total suspended solids (TSS) increases by about 50 % compared to an upstream location (Chapter 2). Suspended solids mainly composed of cellulose fibres blanket the riverbed and exert an oxygen demand up to double that which may be exerted by municipal sludge whilst soluble organics cause depletion of oxygen in the receiving stream (Eckenfelder, 1980; 2000). The wastewater from PANPAPER Mills is thus, a priority environmental issue within the Lake Victoria basin.

Natural and constructed wetland systems remove many pollutants including organic matter and suspended solids (Cooper *et al.*, 1996; Kadlec and Knight, 1996; Vymazal, 2001). Published literature on the use of constructed wetlands for the treatment of wastes from paper manufacturing has concentrated on pulp mill effluents and not effluents from paper mills, e.g. Thut (1989; 1993) and Moore *et al.* (1994). There are no data published for tropical climatic conditions. PANPAPER Mill's effluent, however, incorporates paper mill wastewater and effluent from the manufacture of chlorine gas using electrolysis of brine, and a recycling plant for waste paper.

In reported studies both tertiary and secondary constructed wetland treatment systems are used but efficiency is variable: biochemical oxygen demand (BOD) removal efficiency ranges from 6-90% and TSS from 33-81%; (Thut 1989 & 1993; Hammer *et al.*, 1993, Moore *et al.*, 1994). Preliminary surface flow wetland mesocosm studies conducted in 1999 - 2000 by Abira *et al.* (2003) in Kenya indicated that emergent macrophytes are suitable for pulp and paper mill wastewater improvement provided appropriate operation conditions are applied. Removal of COD was found to be in the range 10 – 55% and 15 – 65 % at five- and ten-day hydraulic retention times respectively. TSS removal was in the range 44 – 86% and 21 - 66 % respectively for the two retention times. The study did not include regular measurements of BOD.

Processes and influencing factors

In constructed wetlands the reduction of organic matter and suspended solids is accomplished by diverse treatment mechanisms namely: sedimentation, filtration, adsorption, microbial interactions, and uptake by vegetation (Watson *et al.*, 1989; Cooper *et al.*, 1996; Kadlec *et al.*, 2000). Besides the inflow and outflow, the solids mass balance is mainly influenced by filtration, plant and microbial growth and decay as well as the prevailing hydrological and hydraulic conditions. Dissolved organic matter is removed mainly by microbial degradation and to a lesser extent, by uptake from plants.

Reduction of organic matter follows first order kinetics. Considering *in-situ* wetland generation of BOD (a non-zero background BOD) the reaction is represented as follows (Kadlec and Knight, 1996).

$$\ln\frac{\left(C_o-C^*\right)}{\left(C_i-C^*\right)}=-\frac{k}{q}$$ Equation 5.1

Where:
C_o = mean outflow pollutant concentration, g/m^3
C_i = mean pollutant concentration at start, g/m^3
C^* = non-zero background BOD (g/m^3)
k = reaction rate constant, m/d
q = hydraulic loading rate, m/yr

The rate of BOD generation by a wetland, r (g/m^2.day) is given by:

$$r = kC^*$$ Equation 5.2

In order to biologically treat pulp and paper mill wastewater successfully, all treatment wetlands must be supplemented with nitrogen and phosphorus since they are deficient in these microbial nutrients essential for high rates of oxidation (Edde, 1984; Gehm and Bregman, 1976). The acceptable BOD_5: nitrogen: phosphorus ratio in a biological treatment system is usually considered to be 100:5:1 (Water Pollution Control Directorate, 1976). However, the actual nutrient requirement is a function of temperature and the growth phase of the microorganisms. At PANPAPER Mills clarified effluent is dosed with di-ammonium phosphate fertilizer (150 kg/day). The BOD:N:P ratio in the final effluent is 100:5.7:1.7 (Chapter 2). This ratio indicates nitrogen limitation.

Under the Lake Victoria Environment Management Project, a pilot-scale study was initiated to establish the efficacy of a constructed wetland in removing organic matter and suspended solids from pre-treated pulp and paper mill wastewater and to recommend optimal design criteria and operating conditions for a full-scale constructed wetland.

The main objectives of this study were:

1. To establish the purification efficiency with respect to BOD, COD and TSS in a constructed wetland under varying hydraulic loading rates, macrophyte growth conditions and operation modes.
2. To determine the reaction rate parameters and residuals for organic matter (BOD) removal in the constructed wetland.

Materials and Methods

This research study was conducted between November 2002 and August 2005 at the premises of PANPAPER Mills. The design, layout, hydrology and flow characteristics of the constructed wetland were as described in Chapter 3.

Experimental approach

The following approach was adopted:

1. Initially the batch load-drain operation mode with 5-day and 3-day cycles was undertaken. It was assumed that this format would provide alternate aerobic-anaerobic conditions that would enhance organic matter degradation (Stein et al., 2003). Initial results indicated that much of the COD present as particulate matter was discharged

with the wetland outflow. It was therefore decided to use continuous flow operation mode.

2. The vitality of the wetland plants was relatively poor at 5-day hydraulic retention time (HRT) (Chapter 4). This was attributed to nitrogen limitation in the wetland and to additional stress that was caused by the draining of the wetland during batch operation. A shorter retention time of 3 days was therefore used.

Experimental set up

Batch loading operation
This was carried out using subsurface flow wetland cells (Series A-D) in three different phases from November 2002 to March 2003 (phase 1, 5-days HRT), March 2003 to July 2004 (phase 2, 3-days HRT) and from February to April 2005 (phase 3, 5-days HRT). Phase 3 was a repeat of phase 1 when the wetland system was considered to be mature. *Typha* cells (series B) were operated for only 25 days. The wetland was fed with wastewater from the final stabilisation pond.

Continuous flow operation
The SSF wetland cells (series A-D) were initially operated at a mean hydraulic loading rate (HLR) of 4.1 - 4.9 cm/day while the FWS cells (Series E) were operated at HLR of 9.3 cm/day for 325 days (first phase - March 2004 to February 2005) with wastewater from the final stabilisation pond. Plant shoots were harvested after 270 days (December 2004, corresponding to day 800 from start of study) to remove mature and senescing shoots and allow re-growth of new ones. In the second phase all wetland cells (series A-E) were operated as SSF systems with mean HLR of 4.9 – 5.7 cm/day (series A-D) and 9.8 cm/day (series E) from April - August 2005. The objective was to establish the performance of the mature wetland when plant shoots were at an active growth stage. *Typha* cells (series B) were loaded with wastewater from the second aeration pond of PANPAPER Mills in order to determine how far pre-treatment is necessary.

Field measurements, sampling and analysis

On-site parameters that were monitored included pH, electrical conductivity (EC), dissolved oxygen (DO), redox potential (Eh) and temperature using a portable Multi-parameter Water Quality Monitor, model 6820-10M-0, manufactured by YSI Incorporated, USA.

Wastewater samples (wetland cell inflow and outflow) were collected every two to four weeks for BOD, COD, and TSS analysis. During batch operation sampling was at the beginning and the end of each batch cycle. The samples were transported to the laboratory, and analysed according to *Standard Methods* (APHA, 1995).

Precipitation data were collected using a rain gauge installed on site. Evaporation and evapotranspiration during batch loading were determined from topping up water on a daily basis. Hence the outflow was assumed to be equal to inflow. For continuous flow, inflow and outflow rates were measured twice daily (Chapter 3). During the batch loading operation the wetland cells were calibrated as described in Chapter 3.

Determination of BOD reduction rates and wetland residuals

Equation 5.1 was used to calculate the areal reaction rate constant k (m/yr). Corrections were made for precipitation and evapotranspiration in Equation 5.1 as recommended by Kadlec (1997). Wetland BOD residuals and BOD generation rates were calculated using Equations 5.1 and 5.2 respectively.

Data Analysis

Means were compared using appropriate tests in SPSS statistical software after testing for normality and homogeneity. For normally distributed data, the appropriate parametric tests such as repeated measures ANOVA were performed. Multiple comparisons were performed using Tukey's HSD tests. Data that were not normally distributed even after log transformations were analysed using appropriate non-parametric tests. Independent samples were performed by the Mann U Whitney or Kolmogorov-Smirnoff tests as appropriate. Comparison of in-out data for determination of mass removal efficiencies under the continuous flow mode took into consideration the time lag in the wetland. Outflow data were matched with mean inflow data of the previous 3 – 5 days.

Results

Wetland inflow wastewater characteristics

The characteristics of the wastewater fed into the constructed wetland are presented in Table 5.1. Wetland inflow wastewater quality was highly variable with little or no dissolved oxygen. COD concentration was approximately 10 times that of BOD. The COD was mainly in the dissolved form (59 - 81 %). Organic matter (BOD and COD) and TSS concentrations were highly variable due to changes in factory operations and de-sludging of treatment ponds.

On-site parameters

The mean pH of outflow from various planted cells showed a reduction of between half to one pH unit when compared to inflow but remained circum-neutral (Table 5.2). The pH for the control cells remained relatively the same as that of the inflow. Wetland dissolved oxygen concentrations were low (less than 2 mg/l) throughout the study. The redox potential (Eh) was indicative of anoxic and anaerobic reducing environments especially during the continuous flow phases. The Eh of the control cells was significantly lower than that of the planted cells.

Table 5.1 Characteristics of the wetland inflow in batch (phases 1–3) and continuous flow (phases 1&2)

		Batch			Continuous flow	
		phase1	phase2	phase3	phase1	phase2
pH	range	7.0 - 8.3	6.9 - 8.0	7.8 - 8.0	7.3 - 8.3	7.9 - 8.3
	Mean/se	7.6±0.16	7.8±0.07	7.9±0.01	8.0±0.05	8.1±0.09
	n	10	15	17	39	5
DO mg/l	range	0 - 1.45	0.12 - 1.9	0.0 - 0.65	0 - 2.8	0 - 1.7
	Mean/se	0.45±0.19	0.6±0.11	0.25±0.07	0.42±0.13	0.51±0.35
	n	10	14	11	28	
EC µS/cm	range	1673 - 2167	1296 - 2650	1580 - 1660	899 – 1851	1358 – 1586
	Mean/se	1920±62	1640±83	1618±7	1567±25	1496±43
	n	10	17	14	36	5
Temperature °C	range	23.4 - 30.5	21.7 - 28.5	24- 27.5	23.5 - 29.6	23.8 - 27.2
	Mean/se	25.6±0.54	25.3±0.51	25.7±0.24	26.0±0.25	25.3±0.54
	n	10	17	17	36	5
BOD mg/l	range	15 - 59	21 - 54	22 - 63	27– 56 (120)*	26 - 48
	Mean/se	38.3±5.6	35±3.1	36±2.2	38.9±1.6	33.9±2.3
	n	10	10	20	27	9
COD mg/l (total)	range	213 - 617	160 - 415	185 - 514	302 - 634 (1115)*	245 - 359
	Mean/se	364±21	296±8.5	354±17	440±9.5	304±12
	n	23	22	21	30	11
COD mg/l (dissolved)	range	232 - 331	230 - 248	146 - 378	248 – 441	168 - 320
	Mean/se	277±9.7	240	284±13	341±18.5	238±14
	n	12	2	16	10	11
TSS mg/l	range	20 - 71	22 - 91	35 - 92	27 - 114 (200)*	33 - 61
	mean/se	48.6±2.7	58±3.7	57.3±4.3	70±3.92	45±3.9
	n	28	22	21	35	12

The data for batch operation were collected from November 2002 to July 2003, July 2003 to March 2004, and February to April 2005 respectively while that for the continuous flow mode were collected from March 2004 to January 2005 and April 2005 to August 2005 respectively. * High pulse of BOD, COD, and TSS caused by de-sludging of PANPAPER Mills' lagoons in July-August 2004.

Table 5.2 In situ measurements during various operation modes. n = 14, 26, 19, 50, and 6 for batch phases 1- 3 and continuous flow (c'flow) phases 1 - 2 respectively except for *Cyperus immensus* where n = 8

Operation Mode and phase		Parameter	Wetlands cells				
			Control	*Typha*	*Phragmites*	papyrus	*Cyperus immensus*
Batch loading	1	pH	-	7.6±0.07	7.6±0.07	7.7±0.05	7.6±0.11
		EC µS/cm	-	1902±43	1715±54	1751±57	1799±26
		DO mg/l	-	0.52±0.14	0.37±0.08	0.39±0.09	0.28±0.12
		Eh mV	-	n.d	n.d	n.d	n.d
	2	pH	7.7±0.04	7.5±0.05	7.5±0.06	7.6±0.04	-
		EC µS/cm	1535±26	1560±45	1577±50	1554±80	-
		DO mg/l	0.86±0.08	1.2±0.13	1.2±0.18	1.1±0.13	-
		Eh mV[1]	24.1	47	61	89	-
	3	pH	7.5±0.02	6.9±0.01	6.9±0.02	7.2±0.02	-
		EC µS/cm	1614±6	2029±27	1880±30	1731±30	-
		DO mg/l	0.31±0.09	0.61±0.16	0.39±0.09	0.32±0.09	-
		Eh mV	-24±4.4	12±0.9	8.0±1.6	-7.1±1.4	-
			Control	*Typha*	*Phragmites*	papyrus	*Typha*[2]
C'flow	1	pH	8.0±0.06	7.4±0.06	7.5±0.06	7.7±0.05	7.5±0.05
		EC µS/cm	1453±30	1595±25	1642±28	1512±21	1536±25
		DO mg/l	0.84±0.13	0.66±0.12	0.58±0.10	0.53±0.09	0.88±0.12
		Eh mV	-59±3.7	-26±3.1	-31±3.2	-43±2.8	-31±2.5
	2	pH	7.8±0.06	7.2±.04	7.3±0.04	7.4±0.06	7.3±0.06
		EC µS/cm	1438±29	1584±76	1540±44	1414±36	1435±57
		DO mg/l	0.04±0.03	0.1±0.05	0.0	0.0	0.18±0.06
		Eh mV	-42±3.8	-8.8±2.1	-17±3.8	-23±1.9	-16±3.8

Eh = redox potential, [1]the data are average of 2 measurements, n.d = not done. *Typha*[2] = Series E cells operated as free water surface in continuous flow phase 1 and as subsurface system in continuous flow phase 2.

COD

Total COD removal was low throughout batch operation mode when compared to the continuous flow (Table 5.3). The mean removal efficiency for the subsurface flow cells was in the range of 22 – 30 % and 39 – 52 %, respectively for the two modes of operation. Removal was variable in both modes and was especially low during periods of storms. There was no significant difference (p> 0.05) between planted and control cells and between cells with different plant species in batch operation.

Table 5.3 COD (total) removal efficiency and wetland outflow concentration in the three phases of wetland batch operation (n = 30, 48 and 27, respectively) and continuous flow phases 1&2 (n = 86 and 12, respectively). Similar letters (a - c) imply there is no statistically significant difference in the removal efficiency or outflow concentration at 95% confidence interval.

Cell type	Mean COD removal efficiency (%)			Mean COD outflow mg/l		
	Batch operation					
	Phase 1	Phase 2	Phase3	Phase 1	Phase 2	Phase3
C. immensus[1]	22.0±3.2	-		226±8	-	
C. papyrus	25.2±3.2	29.4±1.8	27.8±4.0	220±7	195±7	202±6
P. mauritianus	30.0±3.2	27.4±1.5	23±4.7	224±9	204±7	217±8
T. domingensis	24.5±3.3	29±1.3	39±10	225±5	199±6	234±26
Control (unplanted)	-	29.3±1.2	33±2.7	-	201±7	212±9
	Continuous flow					
	Phase 1	Phase 2		Phase 1	Phase 2	
C. papyrus	39.2±1.3 [b]	37.4±2.0 [ab]		335±6 [b]	209±6 [ab]	
P. mauritianus	46.3±1.4 [c]	40.5±2.8 [ab]		353±8 [c]	217±8 [abc]	
T. domingensis	47.1±1.9 [c]	45.0±3.8 [b]		327±8 [b]	219±5 [bc]	
Control (unplanted)	51.6±1.8 [d]	43.1±2.6 [b]		297±8 [a]	201±6 [a]	
Typha[2]	30.1±1.3 [a]	32.9±2.0 [a]		355±8 c	228±7 [c]	

[1] n = 24, [2] Series E cells operated as free water surface in phase 1 and as subsurface flow in phase 2

Multiple comparisons with Tukey's HSD post hoc tests on the continuous flow phase 1 data revealed that the control cells had significantly higher removal efficiencies and lower COD concentration in the outflow. Papyrus cells had the lowest removal efficiency among the planted cells (Series B&C). The higher loaded FWS *Typha* cells (Series E) had a COD outflow concentration similar to that of *Phragmites* cells. COD removal efficiency was similar for all cells (series A-D) in the continuous flow phase 2 during plant re-growth (post harvest). Series E cells with higher influent loading however exhibited similar performance with *Phragmites* and papyrus cells.

Dissolved COD removal (continuous flow) was in the range of 27 – 42 % with the highest removal being in the control cells while the lowest was in the papyrus cells. Series E cells (FWS) had significantly lower mean removal efficiency (21 %). Dissolved COD concentration in the wetland outflow accounted for 90 – 95 % of total COD.

BOD

The removal efficiency of BOD was high throughout the study (Figure 5.1) yielding outflow concentrations less than the new Kenyan guideline value of 30 mg/l in both batch feed and continuous flow operations (Table 5.4). There was no significant difference in the removal efficiency between the two modes of operation. However, outflow BOD concentrations were higher in continuous flow. Outflow BOD concentration from *Typha* series E cells with higher loading (9.3 to 9.8 cm/day and 35 – 60 kg BOD/ha.day) were not significantly different for the two system types (subsurface and free water surface flow). There was no significant difference in outflow BOD concentration between *Typha* cells series B and series E in continuous flow phase 2. The outflow concentrations in batch phase 2 and 3 were not significantly different despite increased removal efficiency for all cells (series A-D). Wetland cells planted with different macrophyte species did not have a significant difference in performance except in continuous flow phase 1 where the control and *Typha* (series B) cells had better removal efficiency than *Phragmites* and papyrus cells. The mean removal efficiency for cells planted with *Cyperus immensus* (74 %) in the first phase of batch feed is unrepresentative due to the few data (n = 4).

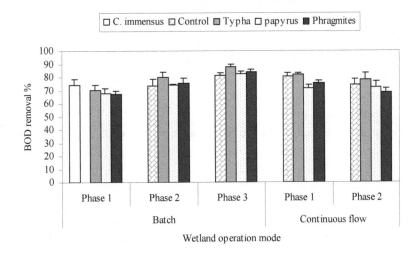

Fig. 5.1 BOD removal efficiency in various phases of wetland operation. n = 10, 8, 27, respectively for batch phases; n = 40 and 10, respectively for continuous flow).

Table 5.4 Mean BOD outflow concentrations from various wetland cells.

Cell type	Mean BOD outflow concentration mg/l				
	Batch operation			Continuous flow	
	Phase 1	Phase 2	Phase3	Phase 1	Phase 2
Papyrus	10.3±2.4	5.6±0.4	4.9±0.4	14.7±0.9	12.6±2.9
Phragmites	9.4±1.8	5.3±0.2	4.7±0.49	15.3±1.0	15.1±1.8
Typha	11.4±3.6	4.4±0.4	4.4±0.9	10.3±0.6	11.3±2.9
Control (unplanted)	-	5.8±0.4	5.3±0.48	11.1±1.3	12.2±1.9
Typha (FWS – series E)	-	-	-	17.3±1.2	-
Typha (SSF – series E)	-	-	-	-	14.4±3.9

n was variable (10, 8, 27, respectively for the batch phases; n = 40 and 10, respectively for continuous flow).

TSS

Removal efficiencies were lower in batch operation compared to continuous flow. Mean removal efficiencies for various wetland cells (Series A-D) in batch loading were in the range of 68 – 76 % (phase 1), 66 - 68 % (phase 2) and 35 - 71 % (phase 3). In batch phase 3 the control cells had significantly higher removal efficiency (71 %) than the planted cells (35 - 62 %). The mean removal efficiency for continuous flow was in the range of 80 – 85 % and 76 – 84 % in phases 1 and 2 respectively. Mean outflow TSS concentrations were in the range of 8 - 21 mg/l (Table 5.5) which is less than the Kenya guideline limit of 30 mg/l except for *Phragmites* and *Typha* cells in batch phase 3 operation (32 and 31 mg/l respectively). In the continuous flow mode, *Typha*-planted cells had consistently higher removal efficiency than *Phragmites* and papyrus cells. However, the performance of *Typha* cells when compared to that of the control cells was similar in phase 1 but significantly higher ($p<0.05$) in phase 2. Series E cells with higher hydraulic loading had wetland outflow TSS of 14 mg/l when operated as subsurface flow. Outflow TSS concentration was significantly lower ($p< 0.05$) in phase 2 than in phase 1 of continuous flow for all cells.

Table 5.5 Mean TSS outflow concentrations from various wetland cells. Series E cells received double the TSS loading of the other cells. Similar letters (a-c) imply there is no statistically significant difference in removal efficiency or outflow concentration at 95% confidence interval.

Cell type	Mean TSS outflow concentration mg/l				
	Batch operation			Continuous flow	
	Phase 1	Phase 2	Phase3	Phase 1	Phase 2
Papyrus	11.5±2.4	15.5±0.95	17.2±0.85	19.8±1.0 [ab]	10.4±0.76 [ab]
Phragmites	15.5±3.7	14.9±0.61	32±2.2	21.4±1.4 [b]	10.8±0.97 [ab]
Typha[1]	8.5±0.83	16.6±1.2	31±5.6	17.7±0.6 [ab]	8.2±0.86 [a]
Control (unplanted)	-	15.5±1.1	14.0±0.87	16.1±1.0 [a]	12.0±2.4 [bc]
Typha (FWS) series E	-	-	-	27.4±1.3 [c]	-
Typha (SSF) series E	-	-	-	-	14.1±1.3 [c]

n was variable: 16, 26, 28, respectively for batch phases; n = 60 and 12, respectively for continuous flow; for *Typha*[1] n = 8

BOD reduction rates, background concentrations and internal generation rates

Equation 5.1 was used to calculate areal rate constants (k) for BOD when the wetland was in a mature steady growth stage (continuous flow phase1). These were compared with rates during the plant exponential growth stage in phase 2 (Table 5.6). k values were higher at higher hydraulic loading rate (series E). The non-zero background (residual) concentration and the wetland BOD generation rates were derived from Equations 5.1 and 5.2 and are also presented in Table 5.6. *Phragmites* cells had the highest residual BOD concentrations among the series A-D cells. *Typha* series E cells that were loaded at a higher rate had higher BOD residual than the corresponding series B cells. BOD generation rates were also higher in the series E cells.

Table 5.6 BOD areal rate constants, residuals and generation rate for the continuous flow mode.

Phase	System type	Plant species	Residual BOD (C*) mg/l	BOD areal rate constant (k) m/d	BOD generation rate (r) g/m².day
1	SSF	Control - unplanted	4.58	0.061	0.277
1	SSF	*Phragmites*	7.43	0.058	0.432
1	SSF	Papyrus	5.29	0.055	0.291
1	SSF	*Typha*	4.44	0.069	0.308
1	FWS	*Typha* (Series E)	6.17	0.093	0.576
2	SSF	Control - unplanted	4.58	0.073	0.333
2	SSF	*Phragmites*	6.65	0.060	0.400
2	SSF	Papyrus	4.32	0.069	0.297
2	SSF	*Typha*	4.93	0.065	0.325
2	SSF	*Typha* (Series E)	6.44	0.114	0.734

Series E cells were loaded at the rate of 9.3 cm/day and 9.8 cm/day in phases 1 and 2, respectively while the rest were loaded at 4.1 – 4.9 cm/day and 4.9 – 5.7 cm/day, respectively. Mean BOD inflow concentration for phases 1 and 2 were 38.8±1.6 and 33.9±2.3 respectively.

Discussion

Ambient conditions

The water temperature of the wetland followed the prevailing air temperature as discussed in Chapter 3. The pH of the wetland outflow remained circum-neutral albeit with a lowering in the planted cells. Many treatment wetlands exhibit a strong buffer capacity with respect to pH (Kadlec and Knight, 1996). The difference in pH between planted and unplanted cells may be due to decomposition of organic matter of wetland plant origin. Dissolved oxygen remained low in the wetland resulting in a strong anaerobic and reducing environment as indicated by the redox potential. The control cells exhibited lower redox potential than planted cells an indication that plants contributed to oxygenation of the wetland beds (Brix, 1997). Due to low dissolved oxygen it is assumed that anoxic and anaerobic degradation processes were predominant (Kadlec *et al.*, 2000).

Purification efficiency

The BOD and TSS removal efficiencies achieved in this study are within the upper range of 25 – 90 % and 33 - 81 %, respectively reported in literature for wetland systems receiving pulp mill wastewater (Thut, 1993; Hammer *et al.*, 1993; Tettleton *et al.*, 1993 and Moore *et al.*, 1994). The outflow BOD and TSS were within the Kenyan guidelines of 30 mg/l each. Data on COD removal in pulp and paper mill treatment by constructed wetlands are lacking in the literature. However, COD removal efficiency found in this study (22 – 52 %) is comparable to those reported for a wetland bucket mesocosm study by Abira *et al.* (2003). Masbough *et al.* (2005) reported 25 - 51% and 42 % removal respectively for COD and lignin in a wetland receiving wood waste leachate in Canada. Low removal of COD is attributed to the presence of cellulose fibres, lignins and lignin degradation products in pulp and paper mill wastewater (Edde, 1984). These do not contribute to the immediate oxygen demand, as they are not readily biodegradable. The proportion of BOD removed therefore was high compared to that of COD. In general the removal efficiencies for organic matter and suspended solids was variable and seemed to depend on system operation/loading mode, hydraulic loading rate, retention time and presence or absence of plants.

Influence of operation mode
COD and TSS removal efficiency were significantly higher in the continuous flow compared to the batch-loading mode. BOD removal, however, was not significantly different between the two modes of operation. The load-drain batch mode was initially adopted with the assumption that it would aerate the wetland bed and enhance organic matter degradation (e.g. Stein *et al.*, 2003). Although some aeration appears to have occurred as indicated by higher redox potentials, there was no concomitant increase in removal efficiency for COD. The lower efficiency in removal for both COD and TSS may be attributed to resuspension of previously deposited solids in the gravel voids. Kadlec and Knight (1996) described TSS outflow loads from subsurface flow wetland systems as reflecting a balance between removal, resuspension and generation.

Influence of Hydraulic retention time/loading rate
There was no significant difference in the removal efficiency of COD and TSS between batch phases 2 and 3 with three and five-day retention times respectively (Tables 5.3-5.5) except for TSS removal in *Phragmites* and *Typha* cells in phase 3. BOD removal was significantly higher at the longer retention time although the outflow concentrations remained similar. This indicates that maximum organic matter removal was achieved within 3 days and additional retention time did not improve wetland effluent quality. This may be due on one hand to internal generation processes characteristic of wetlands that result in non-zero background concentrations and to the resuspension by the load drain format on the other.

The wetland displayed good buffering capacity with respect to organic matter and TSS. This was indicated by similarity in outflow concentrations between cells with higher hydraulic loading rate (series E) and those with lower loading rate (series A-D) in the continuous flow phase 2 operation (Table 5.3-5.5).

Influence of plants
In the continuous flow phase 1 operation COD removal efficiency was significantly higher in unplanted compared to planted cells (Series A-D) while in phase 2 (post harvest) there was no significant difference. BOD removal was not influenced by presence or absence of plants. However, both *Typha* (series B) cells and the controls had significantly higher removal efficiency than other cells in continuous flow phase 1. With regard to TSS, wetland outflow concentrations were significantly lower in phase 1 when plants were at steady growth stage than in phase 2 during active growth (post harvest). There was no significant difference between planted and unplanted cells. Only *Typha* series B cells performed better than unplanted cells in phase 2.

Organic matter and suspended solids balance in wetlands is dependent upon the inflow and outflow concentrations as well as generation for example through rhizo-deposition, litter leaching, mineralisation, microbial growth and decay, and resuspension (Kadlec and Knight, 1996). In the continuous flow phase 1 the planted cells were at a steady growth stage with some senescing shoots and litter deposited on the gravel surface (Chapter 4). Decomposition of litter and the senescing plant parts release organic matter that is not readily degradable, and particulate matter. This process may have resulted in the lower removal efficiency of COD by planted cells and higher TSS concentrations in continuous flow phase 1 compared to phase 2.

Typha cells had consistently higher TSS removal efficiency than *Phragmites* and papyrus in continuous flow. This may be due to differences in plant root penetration depth (Chapter 4).

Typha roots penetrated the entire bed depth (approx. 30 cm) while Phragmites and papyrus rooting depths were at the top 20 cm and 10 cm, respectively. Better filtration was therefore achievable in *Typha* cells.

BOD reduction rates and background concentrations

Areal BOD reaction rate constants varying from 0.055 – 0.114 m/day (20 – 42 m/yr), obtained in this study, are reported for pulp and paper mill wastewater for the first time. The rates however, are comparable to those reported in literature for wetland systems receiving domestic and municipal sewage. Okurut (2000) obtained rates between 0.039 – 0.099 in a free water surface flow (FWS) constructed wetland receiving presettled municipal sewage in Jinja, Uganda. Other reported rate constants for subsurface flow wetlands in various parts of the world (mainly temperate climate) are variable and in the range of 0.06 – 1.0 m/day depending on such factors as hydraulic loading rates, inflow concentration and water depth (Kadlec, 1997, 2000).

The residual BOD concentrations (C^*) found in this study varied between 4.3 and 7.4 mg/l (Table 5.6). Okurut (2000) reported residuals between 12 and 19 mg/l for the Jinja wetland while Kadlec and Knight (1996) report an average of 6 mg/l and 9 mg/l for FWS and SSF wetlands respectively. C^* values are weakly dependent on inflow BOD concentrations but is influenced by inflow nutrient amounts that are responsible for the biochemical cycle. Higher loading of *Typha* cells (series E) in this study may therefore be responsible for the higher BOD residual. *Phragmites* cells had the highest BOD residual in phase 1 compared to similarly loaded cells (series, A-D). This may be due to larger amount of litter from leaf fall (Chapter 4) on the gravel surface of the *Phragmites* cells.

Implications for Nzoia River water quality

The current discharge of wastewater from PANPAPER Mills' treatment ponds into the River Nzoia results in an increase in downstream (3 km) concentrations of 45 %, 75 % and 120 % for TSS, COD and BOD, respectively compared to an upstream location (Chapter 2). Integration of a constructed wetland of similar purification efficiency as in this study (as a tertiary treatment stage) would result in lower outflow concentrations for organic matter and TSS as presented in Table 5.7.

It is noteworthy that even at higher loading rates for example as with constructed wetland cells Series E (phase 2) predicted outflow concentrations are comparable to those with lower loading rates. Therefore if a wetland with a similar loading rate and performance were integrated with PANPAPER Mills treatment system, downstream pollution would be reduced from the current concentrations by 44 %, 30 % and 63 % for TSS, COD and BOD (Figure 5.2).

Table 5.7 Water quality of River Nzoia measured upstream and downstream of PANPAPER Mills's discharge and predicted concentrations for various scenarios of constructed wetlands (CW) integrated with existing treatment ponds.

Scenario		Concentration mg/l			% Reduction in pollution from the current level		
		TSS	COD	BOD	TSS	COD	BOD
River Nzoia measured	Upstream point	24	20	1			
	Downstream (DS) point ("business-as-usual")	35	35	2,2			
River Nzoia predicted	DS with CW (phase 1) & *Typha* Series B cells	24,7	32,9	1,4	43	10	79
	DS with CW (Phase 1) & *Typha* series E cells	25,1	34,1	1,7	41	5	50
	DS with CW (Phase 2) & *Typha* Series B cells	24,3	28,7	1,4	45	32	75
	DS with CW (Phase 2) & *Typha* Series E cells	24,6	29,0	1,6	44	30	63

Upstream = 500 m before PANPAPER Mill effluent discharge point. Downstream = 3 km after discharge point

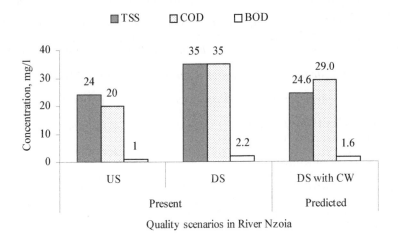

Fig. 5.2 Predicted TSS and organic matter concentrations downstream of effluent discharge point if a constructed wetland is integrated as tertiary treatment stage with PANPAPER Mills's ponds. US = upstream sampling point, DS = downstream sampling point, and CW = constructed wetland with *Typha domingensis* and hydraulic loading rate of 9.8 cm/day.

Implications for design and operation

The BOD reaction rate constant obtained in this study for *Typha* series E (0.114 m/day or 42 m/yr) is higher than that obtained for phenols (ca 36 m/yr) in the same wetland cells (Chapter 6). Most wetlands' design is based on BOD removal. However, Kadlec and Knight (1996) recommend wetland sizing on the regulated parameter requiring the largest area (with a lower reaction rate constant). In the case of wastewater from PANPAPER Mills the size recommended for phenol removal in Chapter 6 (approx. 35 ha) would be appropriate for BOD removal as well.

Wetland outflow quality from cells operated at higher hydraulic loading rates was comparable to similar cells operated at about half the rate. The higher rate (average 9.8 cm/day) would hence be appropriate for loading of a full-scale wetland.

Conclusion

The constructed wetland in this study effectively removed BOD and TSS from the pre-treated pulp and paper mill wastewater to concentrations below that prescribed by the regulating authority in Kenya. However, the guideline of 100 mg/l for COD was not achieved in this study. Notwithstanding, integrating a full-scale constructed wetland with the current treatment system of PANPAPER Mills significantly improves the effluent quality and would reduce pollution in the Nzoia river from current levels by 30 %, 44 % and 63 % for COD, TSS and BOD respectively.

References

Abira, M. A., Ngirigacha, H. W. and van Bruggen J.J.A. 2003. Preliminary investigation of the potential of four tropical emergent macrophytes for treatment of pre-treated pulp and papermill wastewater in Kenya, *Water Science and Technology* 48 (5): 223 - 231

APHA, 1995. *Standard methods for the analysis of water and wastewater*, 19th edition. American Public Health Association, Washington DC.

Brix, H., 1997. Do macrophytes play a role in constructed treatment wetlands? *Water Science and Technology* 35 (5): 11 - 17.

Cooper, P.F., Job, G.D., Green, M.B., and Shuttes, R.B.E. 1996. *Reed beds and constructed wetlands for wastewater treatment*. WRc plc. Swindon, Wiltshire, U.K., 184, pp.

Eckenfelder, W.W. 1980. *Principles of water quality management*, CBI publishing. Boston, Massachusetts, 717 pp.

Eckenfelder, W. W. Jr. (2000). *Industrial Water Pollution Control*, 3rd edition, McGraw-Hill, New York U.S.A.

Edde, H., 1984. *Environmental control for pulp and paper mills*. Noyes Publications, Park ridge, New Jersey, USA. 500 pp.

Gehm H.W. and Bregman, J.I. (eds.) 1976. *Handbook of water resources and pollution control*. Van Nostrand Reinhold Company, New York, 840 pp.

GOK, 1999. *Kenya population census*, Ministry of Planning and National Development, Central Bureau of Statistics, Government of Kenya.

Hammer, D.A., Pullin, B.P., Mc Murry, D.K. and Lee, J.W. 1993. Testing colour removal from pulp mill wastewaters with constructed wetlands. In: Moshiri, G.A. (ed.). *Constructed wetlands for water quality improvement*. CRC Press, Lewis Publishers, Michigan, pp 5 – 19.

JICA/GOK, (1992). *National Water Master Plan, Data Book: Hydrological data.* Japan International Cooperation Agency and Government of Kenya, Ministry of Water Development.

Kadlec, R.H. and Knight, R.L. 1996. *Treatment wetlands*. CRC Press Inc., Lewis Publishers, Boca Raton, Florida, 893 pp.

Kadlec, R.H., 1997. Deterministic and stochastic aspects of constructed wetland performance and design. *Water Science and Technology* 35 (5): 149 – 156.

Kadlec, R.H., Knight, R.L., Vymazal, J., Brix, H., Cooper, P. and Haberl, R. 2000. *Constructed wetlands for pollution control: processes, performance, design, and operation.* IWA specialist group on use of macrophytes in water pollution control. Scientific and technical report No.8, IWA publishing, London, UK, 156 pp.

Masbough, A., Frankowski, K., Hall, K.J. and Duff, S.J.B., 2005. The effectiveness of constructed wetland for treatment of wood waste leachate. *Ecological Engineering*, 25: 552 – 566.

Metcalf and Eddy, Inc., 1991. *Wastewater engineering: treatment disposal and re-use, 3rd edition.* Revised by G. Tchobanoglous and F.L. Burton, McGraw-Hill, New York, USA, 1334 pp

Moore, J.A., Skarda, S.M. and Sherwood, R. 1994. Wetland treatment of pulp and paper mill wastewater. *Water Science and Technology* 29 (4): 241 - 247.

Okurut, T. O. 2000. A pilot study on municipal wastewater treatment using a constructed wetland in Uganda. PhD dissertation, Balkema publishers, Rotterdam, The Netherlands

Stein, O.R., Hook, P.B., Biederman, J.A., Allen, W.C. and Borden, D. J. 2003. Does batch operation enhance oxidation in subsurface constructed wetlands? *Water Science and Technology* 48 (5): 149 - 156

Tettleton, R.P., Howell, F.G. and Reaves, R.P. 1993. Performance of a constructed marsh in the tertiary treatment of bleach Kraft pulp mill effluent: results of a 2-year pilot project. In: Moshiri, G.A. (ed.). *Constructed wetlands for water quality improvement*, CRC Press, Lewis Publishers, Boca Raton, pp 437 - 440.

Thut, R.N. 1989. Utilisation of artificial marshes for the treatment of pulp mill effluents. In: Hammer, D.A. (ed), *Constructed wetlands for wastewater treatment: municipal, industrial, and agricultural*. Lewis Publishers, Chelsea, Michigan, U.S.A., pp 239 - 244.

Thut, R.N., 1993. Feasibility of treating pulp mill effluent with a constructed wetland. In: Moshiri, G.A. (ed.). *Constructed wetlands for water quality improvement*, CRC Press, Lewis Publishers, Boca Raton, pp 441 - 447.

Vymazal, J. 2001. Removal of organics in Czech constructed wetlands with horizontal sub-surface flow. In: Vymazal, J. (ed.), *Transformations of Nutrients and Constructed wetlands*, pp 305 - 327, Buckhuys publishers, Leiden, The Netherlands.

Water Pollution Control Directorate, 1976. *Proceedings of Seminars on water pollution abatement technology in the pulp and paper industry*. Economic and technical review report EPS 3-WP-76-4 Environment Canada and Canadian Pulp and Paper Association, 220 pp.

Watson, J.T., Reed, S.C., Kadlec, R.H., Knight, R.L., and Whitehouse, A.E. 1989. Performance expectations and loading rates for constructed wetlands. In: Hammer D.A. (ed.), *Constructed wetlands for wastewater treatment: municipal, industrial and agricultural*, Lewis Publishers, Chelsea, Michigan, U.S.A., pp. 319 - 351.

Chapter Six

Phenol removal: efficiency, processes and implications

Publications based on this chapter:

Abira, M.A., van Bruggen, J.J.A. and Denny, P. 2005. Potential of a tropical subsurface constructed wetland to remove phenol from pre-treated pulp and papermill wastewater. *Water Science and Technology* 51 (9): 173 - 176.

Phenol removal: efficiency, processes and implications

Abstract
A pilot-scale study was undertaken to establish the efficacy of a constructed wetland in removing phenols from pre-treated pulp and paper mill wastewater prior to discharge into the Nzoia River within the Lake Victoria basin in Kenya. The study determined the purification efficiency and buffering capacity for phenols in subsurface (SSF) and free water surface flow (FWS) constructed wetland cells under varying hydraulic loading rates, aquatic macrophyte growth conditions and operation modes. The role of various processes and reaction rate parameters for phenol reduction were determined.

Mean phenol removal efficiencies ranged from 73 % to 96 %. The wetland's performance indicated good buffering even during the highest inflow phenol concentration of 1.3 mg/l and the highest hydraulic loading rate (HLR) of 9.8 cm/day in the wetland. For batch operation, optimal removal was achieved at 5-day hydraulic retention time. The major processes of phenol removal are microbial breakdown followed by sedimentation/adsorption. Phenol removal was enhanced by wetland age and presence of emergent aquatic macrophytes especially when they were at an exponential growth stage. During this stage plant uptake rates were 5.4 – 12.7 mg phenol/day accounting for 10 – 23 % of phenol removed in SSF cells.

We report for the first time phenol reduction rates (k values) in a constructed wetland. Average k, for SSF cells was 27 m/yr and 44 m/yr respectively at HLRs of 4.1 - 4.9 cm/day and 9.8 cm/day respectively. k was 38 m/yr for the FWS at a HLR of 9.3 cm/day. Mean volumetric reduction rates (k_v values) for batch operation were 0.57 d^{-1} for *Phragmites mauritianus* and *Cyperus papyrus* cells. We predict that by integrating a full-scale constructed wetland of similar performance as the one in this study into the system the current state of pollution in the Nzoia River would be reduced by more than 90 %.

Key words
Phenols, reduction rates, removal efficiency, phenol budgets, macropyhtes, paper and pulp mill wastewater

Introduction

Pan African Paper Mills (E. A.) Ltd. (PANPAPER) discharges its treated effluent into the River Nzoia. The effluent however, does not meet the guidelines for discharge of phenols (0.05 mg/l) into surface water in Kenya. The river is important for artisanal fisheries and water supply. Discharge of high concentrations of phenols is therefore undesirable. Phenols are toxic to aquatic life and impart a bad taste to chlorinated drinking water at very low concentrations. Biochemical and histological effects on fish and invertebrates in streams receiving pulp and paper mill wastewater have been attributed to chlorinated phenolics, resin and fatty acids extracted from wood during pulping (Thut, 1993). Significantly, head deformities in chironomid larvae attributed to toxic substances that include phenols have been reported in 68 % of a population sampled downstream of the PANPAPER Mills effluent discharge when compared to 8 % in an upstream location (Mogere, 2000).

Constructed wetlands have been used as cost-effective alternatives for domestic and industrial wastewater treatment. They are capable of removing toxic substances such as phenols from wastewater emanating from industries and leachates from municipal landfills (Kadlec and Knight, 1996; Eckhardt et al., 1999; Thut, 1993; Masbough et al., 2005). Polprasert et al. (1996) reported phenol removal in a FWS pilot constructed wetland fed with synthetic wastewater containing pure phenol. Reported studies are few and data that can guide design and operation are still lacking.

This pilot-scale study was undertaken to determine the efficacy of a constructed wetland in removing phenols from pre-treated pulp and paper mill wastewater and to recommend optimal design criteria and operating conditions for a full-scale constructed wetland. An initial study by Abira et al. (2005) in the first 15 months of operation reported encouraging results on the capacity of the constructed wetland to remove phenols under batch loading operation at varying hydraulic retention times (HRT). Mean removal efficiencies were found to be about 60 % and 75 % while outflow phenol concentrations were 0.2 mg/l and 0.11 mg/l for 5-day and 3-day HRT respectively. Contrary to expectation, removal efficiencies were lower at the longer retention time, and the wetland's relatively short age (initial 8 months) was postulated as a reason for the discrepancy. Initial results of a study on the removal of organic matter and suspended solids (Chapter 5) indicated that some of the organic matter present in particulate form was discharged with the wetland outflow. It was thought that some phenols might also have been discharged with the particulate matter. This study determined the performance of the constructed wetland in removing phenols under continuous flow operation. The batch-loading operation at 5-day HRT was repeated in this wetland system in the third year when it was mature.

Processes

The structure of phenolic compounds in pulp and paper mill wastewater is similar to the basic structural units of lignin. Due to the use of chlorine for bleaching of pulp, chlorinated derivatives are also present. Despite their structure, some microorganisms (bacteria and fungi) are capable of degrading them (e.g. Fountoulakis et al., 2002; Banerjee, 1997). In wetlands, phenol removal can be attributed to four main processes, namely; bio-degradation, sorption, plant uptake and volatilisation. The primary mechanism is biodegradation (Eckenfelder, 2000). Polprasert et al. (1996) attributed a 30 % removal to volatilisation, and the rest to other processes in a FWS wetland receiving phenol concentrations of 25 mg/l – 400 mg/l. Although uptake by various plant tissues was reported, the individual system processes such as bio-degradation and adsorption were not disaggregated. Data gathered in the "black box" approach in the wetland under their study is not adequate for elucidating the removal processes in the system in general and the role of the individual system components in particular. To optimise phenol removal in a constructed wetland the various removal pathways need to be elucidated.

According to Kadlec and Knight (1996), the pollutant reduction reaction can be described with a first order model, which in the simplest form (assuming zero background pollutant concentration) is given as:

For batch – homogeneous:

$$\frac{C}{Ci} = \exp(-k_v t)$$ Equation 6.1

Where:
C = the remaining pollutant concentration at time t, g/m^3
C_i = pollutant concentration at start, g/m^3
k_v = reaction rate constant, d^{-1}
t   = reaction time, d
For continuous flow:

$$\frac{C}{Ci} = \exp(-Da)$$ Equation 6.2

Where:
Da = Damköhler number
Da = k_vt for first order volumetric reaction
Da = k/q for first order areal reaction
k = rate constant, m/yr
q = hydraulic loading rate, m/yr

Equations 6.1 and 6.2 were used to determine phenol reduction rates in this study.

The objectives of this study were three-fold:
1. To determine the purification efficiency and buffering capacity for phenols of subsurface and free water surface flow constructed wetlands under varying hydraulic loading rates, aquatic macrophyte growth conditions and operation modes
2. To determine the role of various processes in the removal of phenols in the constructed wetland
3. To determine the reaction rate parameters for phenol reduction in a constructed wetland.

Materials and methods

The research study was conducted from March 2004 to August 2005. The design, layout, hydrology and flow characteristics of the constructed wetland were as described in Chapter 3. An initial study using the batch load-drain operation mode with 5-day and 3-day cycles was undertaken from November 2002 to March 2004 (Abira *et al.*, 2005).

Experimental set up

Continuous flow
The wetland system consisted of four pairs of parallel gravel bed horizontal sub-surface flow cells of dimensions 1.2 m (width) x 3.2 m (length) x 0.8 m (depth). Three pairs were planted, each pair having one of three native emergent macrophyte species namely, *Cyperus papyrus*, *Phragmites mauritianus*, and *Typha domingensis*. The fourth pair was left unplanted as a control (Figure 6.1). A free water surface flow system planted with *Typha domingensis* in two cells was operated alongside for comparison. The cells were fed with pre-treated wastewater from the final stabilisation lagoon of PANPAPER Mills (Chapter 2).

The SSF wetland cells (series A-D) were initially operated at a mean HLR of 4.1 - 4.9 cm/day while the FWS cells (Series E) were operated at HLR of 9.3 cm/day for 325 days (first phase - March 2004 to February 2005). Plant shoots were harvested after 270 days (December 2004) to remove mature and senescing shoots and allow re-growth of new shoots. In the second phase all wetland cells (series A-E) were operated as SSF systems with mean HLR of 4.9 – 5.7 cm/day (series A-D) and 9.8 cm/day (series E) from April 2005 until August 2005. The objective was to establish the performance of the mature wetland when plant shoots were at an active growth stage. *Typha* cells (series B) were however, loaded with wastewater from the second aeration pond of PANPAPER Mills in order to establish the level of pre-treatment necessary for optimal removal of phenols in the wetland.

Batch loading, 5 day-HRT
Wetland operation in batch flow mode reported by Abira *et al.* (2005) was repeated in this system when the system was mature. *Typha* cells (series B) were operated for only 25 days. The experiment lasted 75 days between February and April 2005. All cells were loaded with wastewater from the final stabilisation pond.

Fig. 6.1 A view of the constructed wetland showing different cells

Bucket wetland mesocosms set-up and operation
The experiment for the study of various phenol removal processes was carried out in bucket reactors similar in design and capacity to those described in Abira *et al.* (2003). The set-up was a random block design (five blocks) using five system types (treatments) and two hydraulic retention times (three and six days) with two and three replicates respectively (Table 6.1 and Figure 6.2).

Table 6.1 Overview of different treatments (system types and hydraulic retention times) and layout of wetland bucket mesocosms.

Block 1	Block 2	Block 3	Block 4	Block 5
3-day hrt	6-day hrt	3-day hrt	6-day hrt	6-day hrt
W	GF	GP	WP	GF
GP	WP	GF	GP	W
WP	GP	WP	G	GP
GF	W	GP	W	GP
G	G	W	GF	WP

W = wastewater only, GP = gravel planted, WP = plants suspended in wastewater, G = gravel, GF = gravel + biocide (0.2% formalin)

Fig. 6.2 Wetland bucket mesocosms used for study of phenol removal processes.

Buckets with gravel were set up in a similar manner to those used by Roberts (2004) simulating subsurface wetland systems. The buckets containing plants were planted with young *Cyperus papyrus* seedlings collected from a natural wetland, with no rhizomes that could contain stored phenols. The systems were allowed to acclimatize for two months and enable those with plants to develop good root and rhizome coverage.

Each bucket was operated as a batch reactor. Effluent from the final stabilisation lagoon was fed gently through plastic (PVC) feed pipe in order to ensure plug flow conditions. The control buckets received effluent with 0.2 % formalin to inactivate microorganisms. Each day, pre-settled river water was used to compensate for losses due to evaporation. However, for the wastewater-only buckets the procedure used by Polprasert *et al.* (1996) was followed. They were loaded only once at the beginning of each batch cycle. After every three or six days the change in volume due to evaporation and/or precipitation was measured and a sample taken. The bucket contents were discarded and fresh wastewater was then re-loaded. Occasionally, control buckets with gravel and fed with wastewater containing formalin were sampled to check for bacterial count using plate count agar according to *Standard Methods* (APHA, 1995). This was to ascertain that there was no microbiological activity in the buckets.

Field measurements, sampling and analysis
In-situ parameters that were monitored included pH, EC, DO and temperature. These were done using a portable Multi-parameter Water Quality Monitor, model 6820-10M-0, manufactured by YSI Incorporated, USA. The results and are presented in Chapter 5.

Wastewater samples (wetland cell inflow and outflow) were collected bi-weekly for the analysis of phenols. Samples were also taken from side ports along the wetland length (*Phragmites* cells) in June 2004 to determine spatial phenol reduction. During batch operation mode sampling was undertaken at the end of each batch cycle for both wetland buckets and cells. The samples were preserved with sulphuric acid, transported to the laboratory, steam-distilled and analysed by the direct photometric method according to the procedure described in the *Standard Methods* by (APHA, 1995).

Precipitation data were collected from a rain gauge installed on site. Evaporation and evapotranspiration from the buckets experiment were determined from the volume of topping up water added on a daily basis. Hence, the outflow was assumed to be equal to inflow. For the wetland cells, inflow and outflow rates were determined twice daily (Chapter 3). During batch loading operation the wetland cells were calibrated by complete draining then refilling with known volumes of wastewater as described in Chapter 3.

Determination of phenol budgets
Phenol budgets were calculated for the first and second phases of continuous flow loading. The former represented wetland budgets at the plants' steady state growth phase (mature) while the latter represented budgets at the plant shoot exponential growth stage.

The phenol budgets were determined as follows: 1) Total phenol mass flow (mg/day) in the wetland inflow and outflow was calculated by multiplying mean concentrations (mg/m^3) by mean flow rate (m^3/day). The difference between the phenol mass flow in and mass flow out of each wetland cell was taken as the phenol mass removal rate. 2) The difference in % removal efficiency between planted and unplanted cells was assumed to be due to plant uptake (into plant tissue or adsorbed). The difference was then used to determine the mass removal rate (mg/day) due to plants for each cell. 3) Finally, the phenol mass removal rate due to the combination of microbial activity and sedimentation/adsorption was obtained from the difference between the mass removal rate in 1) and the removal rate due to plants in 2).

Determination of phenol reduction rates
Wetland flow characteristics were not discernable from the tracer study in Chapter 3. However, it was assumed that the flow in the wetland beds was predominantly of plug flow reactor type. Equations 6.1 and 6.2 were therefore used to calculate the volumetric reaction rate constant k_v (d^{-1}) and the areal reaction rate constant k (m/yr). Corrections were made for precipitation and evapotranspiration in Equation 6.2 as recommended by Kadlec (1997). The Equation (6.2) was also used to model outflow phenol concentrations from measured inflow concentrations.

Data Analysis

The generated data were analysed using Microsoft Excel analysis toolpak. Means were compared using appropriate tests in SPSS statistical software after testing for normality and homogeneity. For normally distributed data, the appropriate parametric tests such as repeated measures ANOVA were performed. Multiple comparisons were performed using Tukey's HSD tests. Data that were not normally distributed even after log transformations were analysed using appropriate non-parametric tests. In the latter case multiple comparisons were done using the

Median Test after performing a Kruskal-Wallis or Friedman's two-way analysis of variance as appropriate.

Results

Phenol removal

Influence of inflow phenol concentration
Inflow wastewater quality was highly variable with regard to phenol concentrations and ranged between $0.11 - 1.3$ mg/l. Mean concentrations were 0.50 ± 0.05 mg/l and 0.33 ± 0.07 mg/l respectively for the two phases of continuous flow operation. Wetland inflow phenol concentration during batch (5-day) loading was low and in the range of $0.11 - 0.29$ mg/l with a mean of 0.19 ± 0.006 mg/l. Factory operations during this period mainly involved waste paper recycling and no wood pulping. The change in operations also affected the quality of wastewater in the aerated pond. Overflow phenol concentration from the pond averaged 0.26 ± 0.04 mg/l.

The phenol removal efficiency increased with higher phenol concentration in the influent. The increase was logarithmic, reaching a plateau at about 0.6 mg/l phenol in the influent (Fig 6.3). The correlation between inflow phenol concentration and removal efficiency was significant (Eta squared value = 0.72, p = 0.00). The inflow concentration positively influenced areal removal rate ($p < 0.05$) but had no influence on the outflow phenol concentration.

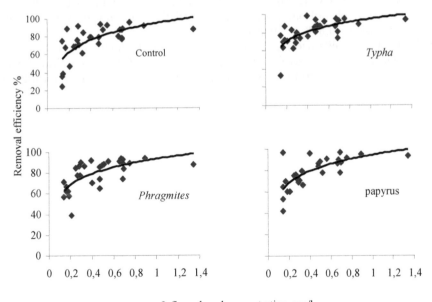

Fig. 6.3 Influence of wetland inflow phenol concentration on the removal efficiency (continuous flow phase 1).

Influence of plants

Continuous flow: Mean removal efficiencies for the first phase of continuous flow operation are presented in Table 6.2. The Kruskal-Wallis test indicated that both plants and system type (subsurface flow or free water surface flow) significantly influenced removal efficiency. The Median Test further indicated that the control unplanted cells had lower removal efficiency during continuous flow phase 1. The difference was however not statistically significant at 95% confidence interval.

Table 6.2 Mean values for phenol removal efficiency and wetland outflow phenol concentrations under continuous flow, first phase: March 2004 to February 2005.

Wetland cell	System type	Removal efficiency, %	Outflow phenol concentration, mg/l	n
Control - unplanted	SSF	82.7±1.5	0.11±0.008	71
Typha domingensis	SSF	83.2±1.6	0.11±0.007	74
Phragmites mauritianus	SSF	83.8±1.2	0.11±0.007	74
Cyperus papyrus	SSF	82.9±1.6	0.10±0.007	70
Typha domingensis	FWS	72.8±1.7	0.15±0.008	76

SSF = subsurface flow, FWS = free water surface flow, n = number of samples

Wetland outflow phenol concentrations were variable ranging from 0.01 to 0.31 mg/l for the SSF cells. There was no significant difference in the mean outflow concentration of the various cells (Table 6.2). The FWS wetland cells however had a higher mean outflow concentration (0.15 mg/l). Outflow phenol had a significant correlation (non-linear) with the areal loading rate (Eta squared value = 0.808, p = 0.00).

Overall, the planted cells had higher removal rates than the unplanted controls (Figure 6.4). Multiple comparisons on log-transformed data with Tukey's HSD test confirmed that the mean areal removal rate for the control was (statistically) significantly lower than for the planted cells (p-value < 0.05).

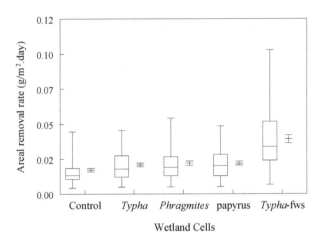

Fig. 6.4 Areal phenol removal rates for various wetland cells during continuous flow phase 1. The bars to the right of each box represent median values (n = 70 – 76), fws =free water surface flow system

In the continuous flow phase 2 with plant re-growth, planted cells had significantly higher removal efficiencies. Phenol removal efficiency averaged 58±4.8 % for the control unplanted wetland cells and 69.5±4.5 % and 72 ±4.8 % for *Phragmites* and papyrus cells respectively. The outflow phenol concentration for the unplanted wetland cells (0.12±0.010 mg/l) was significantly higher (p < 0.05) than for the planted cells: *Phragmites* (0.09±0.009 mg/l) and papyrus (0.07±0.010 mg/l) (Figure 6.5). The mean removal efficiency in series E cells (larger with *Typha*) was 63±6.2 % and mean outflow phenol concentration was 0.09±0.010 mg/l. However, plant re-growth was relatively poor in the series E cells due to attack by vermin monkeys. The mean phenol concentration in the wetland outflow was generally steady over time (Figure 6.6).

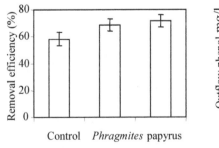
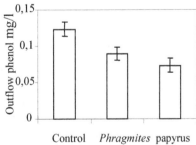

Wetland cells

Fig. 6.5 Phenol outflow concentrations and removal efficiencies in various wetland cells after plant re-growth during continuous flow (n = 13).

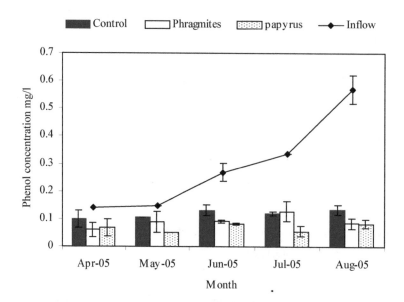

Fig. 6.6 Temporal variations in inflow and outflow phenol concentrations (monthly means) during the second phase (plant re-growth stage) of continuous flow system.

Batch operation. All wetland cells showed good phenol removal efficiency. However, the *Phragmites*-planted wetland cells showed better removal efficiency (95.7±0.7 %) and lower outflow concentration (0.053±0.004 mg/l) than the papyrus cells (92.8±1.4 %, 0.061±0.006 mg/l) and the control (93.4±1.1 %, 0.064±0.005 mg/l). However, the difference was not statistically significant.

Removal mechanisms

Inflow phenol concentration to the wetland bucket mesocosms was variable (0.14 – 0.89 mg/l) with a mean of 0.31±0.025 mg/l. All system types exhibited phenol removal albeit to varying efficiencies (Table 6.3). Phenol removal efficiency was highest (mean up to 71 %) in planted buckets, with or without gravel. Unplanted buckets with gravel had lower removal efficiencies (mean up to 63%). However, comparison of means showed that the difference between planted and unplanted buckets was not statistically significant (p > 0.05). Wetland buckets treated with biocide to inactivate microorganisms, and those with wastewater only, had significantly lower phenol removal efficiencies. There was no statistically significant difference (p> 0.05) in wetland phenol removal efficiency at 3 days compared to 6 days for all system types.

In determining the contribution of various mechanisms to the removal of phenols in the wetlands the following assumptions were made, viz., 1) The removal efficiency in buckets with gravel fed with formalin-containing wastewater represents removal by sedimentation and/or adsorption, 2) The difference in removal efficiency in buckets with either gravel only or gravel with plants was due to biological activity, 3) the difference in removal efficiency in planted buckets (with or without gravel) was due to plant influence (uptake or adsorption onto plant roots).

With the foregoing assumptions it may be deuced that microbial mediated processes had the largest contribution (up to 60 %). Sedimentation and/or adsorption accounted for 30 % while plant uptake accounted for the remaining 10%. However, since plant tissue was not analysed for phenols, the "plant uptake" may represent both active uptake and additional adsorption onto plant roots.

Phenol reduction: reaction rates

Phenol concentration decreased exponentially with distance along the constructed wetland during the continuous flow operation. The variation of phenol concentration along the *Phragmites* cells including a first order trend line is presented in Figure 6.7. Up to 70 % removal occurred within one-quarter of the wetland length. There was no difference in removal between the top upper layer and the lower layer of the wetland bed.

The areal rate constant k (m/yr) was calculated using Equation 6.2 for continuous flow phase 1 (steady state) and with the assumption that there was no internal generation of phenol in the wetland. k was found to be between 26 and 28 m/yr for the planted cell-series B-D. The volumetric reaction rate constant k_v (d^{-1}) for batch operation was 0.53 d^{-1} and 0.6 d^{-1} for papyrus and *Phragmites* cells respectively. The FWS (series E) had a k value of 38 m/yr. The rate constant was higher (44 m/yr) when the series E cells were operated as SSF systems.

Table 6.3 Summary of data for various wetland bucket types: G = gravel, GF = gravel + 0.2% formalin, GP = gravel planted with papyrus, W = water only, WP = planted papyrus in water only. Similar letters (a-d) imply that there is no statistically significant difference in removal efficiency at 95% confidence interval, total cases evaluated = 355.

Retention time (days)	Bucket type	Mean	se	Minimum	Maximum	Median
				% Removal efficiency		
3	G	60.4 cd	4.2	7.1	95.5	64.1
	GF	20.3 a	3.4	6.3	56.6	14.8
	GP	67.4 d	4.4	17.5	97.7	71.3
	W	45.8 bc	3.7	13.9	66.8	51.3
	WP	67.2 d	3.6	9.8	98.3	70.4
6	G	63.0 d	3.0	3.3	96.5	67.2
	GF	29.1 ab	2.9	6.3	55.4	26.4
	GP	71.4 d	2.6	20.9	99.2	77.8
	W	38.1 b	4.1	-9.6	73.8	49.7
	WP	68.2 d	2.9	19.6	99.2	69.4
				Outflow phenol concentration, mg/l		
3	G	0.098	0.010	0.031	0.235	0.087
	GF	0.185	0.020	0.026	0.374	0.170
	GP	0.091	0.011	0.016	0.235	0.089
	W	0.122	0.009	0.041	0.191	0.110
	WP	0.085	0.008	0.007	0.191	0.081
6	G	0.080	0.005	0.015	0.165	0.067
	GF	0.161	0.010	0.075	0.240	0.164
	GP	0.070	0.005	0.003	0.163	0.063
	W	0.130	0.009	0.067	0.265	0.122
	WP	0.077	0.006	0.003	0.204	0.070

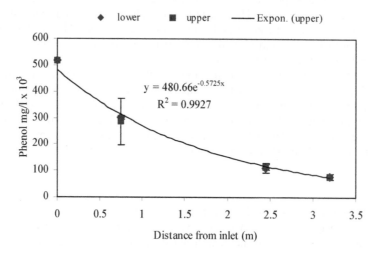

Fig. 6.7 Spatial variations in phenol concentration in Phragmites wetland cells. A first order fit based on upper layer data is indicated.

Simulated outflow phenol concentrations derived from the equation were compared with measured data for *Phragmites* (Figure 6.8). Results of a first order model based on ideal plug flow are also included in the figure for comparison. The measured data were in general low, even during peak inflow concentrations. The model data fairly simulated valley or low inflow concentrations but overestimated outflows during peak inflows.

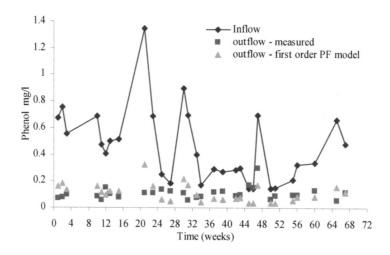

Fig. 6.8 Temporal variations in phenol reduction during continuous flow. Simulated outflow concentrations using plug flow (PF) model with correction for evapotranspiration and precipitation are compared with measured concentrations.

Phenol budgets

Phenol mass balance in the subsurface flow wetland cells is presented in Figure 6.9. During the macrophyte exponential growth stage phenol uptake was high (5.4 – 12.7 mg/day) with *Typha domingensis* having the highest uptake rate while *Phragmites mauritianus* had the lowest. Macrophyte contribution to phenol removal was low during the macrophyte steady growth stage (0.2 – 0.8 mg/day).

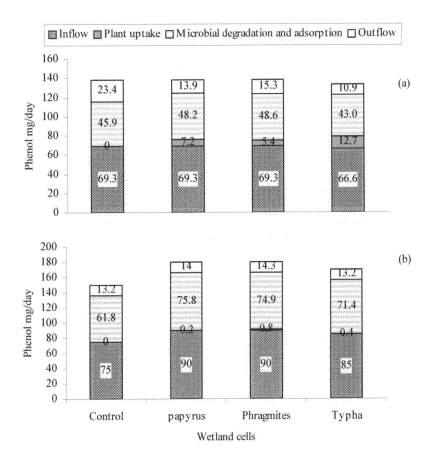

Fig. 6.9. Phenol mass balance in the subsurface constructed wetland cells during macrophyte exponential growth stage, 145 days (a) and at steady growth stage 270 days (b).

Discussion

Purification Efficiency

The constructed wetland effectively removed most of the phenols with mean removal efficiencies ranging from 73 % to 96 %. Polprasert *et al.* (1996) reported similar efficiencies for a FWS wetland receiving synthetic wastewater containing pure phenol concentrations of 25 to 400 mg/l. However, phenol mass removals were higher in their study. The removals in this study are higher than those reported for other treatment systems. For instance, Frigon *et al.*

(2003) reported a 55 % removal of phenols from bark leachate in an aerobic biofilter and 70 % removal by activated sludge. The removal efficiencies obtained in our study, however, were variable and depended on wetland inflow phenol concentration, system type, age, operational (loading) mode, hydraulic loading rate/retention time, the presence or absence of plants, and the growth stage of the macrophytes.

Variations in inflow phenol concentrations were related to factory operations (production, wood type, process changes) and maintenance works on the wastewater treatment lagoons. For example, during partial shutdowns or decreased wood pulping the wetland influent phenol concentration was about 0.3 mg/l while during the desludging of the aerated lagoons, the concentrations were more than 1.3 mg/l.

Phenol removal efficiency was higher in SSF wetland cells when compared to FWS cells with similar macrophytes. This difference was further confirmed by the areal reaction rate constant, k, for series E cells when operated at about the same hydraulic loading rate as SSF and FWS systems. The reaction rates are discussed below. SSF wetlands have a greater substrate surface area for the treatment of contaminants (US EPA, 1993). Although volatilisation is expected to give additional improvement to phenol removal in a FWS system, the plant density in this study shaded the water surface and volatilisation was thought to have had an insignificant role. This was indicated by the small difference in evapotranspiration of the series E cells when operated as FWS and as SSF systems (Chapter 3).

Wetland performance with respect to phenol removal was better when system age was more than two years compared to the performance in the first year. This was seen in the difference in performance between batch 5-day HRT in the first 8 months with mean removal efficiency of 60 % (Abira et al., 2005) and similar operation in the third year during which the efficiency was 93 % – 96 %. The difference is attributed to the more extensive root spread of macrophytes and the time needed for establishment of a microbial environment conducive for bio-degradation of phenol in the wetland.

With the batch-loading format, a higher retention time of 5 days gave a lower phenol concentration in the effluent (0.053 mg/l) when compared to the 3-day retention in the earlier study reported by Abira et al. (2005). Phenol removal efficiency was sustained during continuous flow at higher HLR (9.8 cm/day, series E) yielding outflow concentration similar to that of cells with lower HLR (4.9- 5.7 cm/day, series A-D). This indicates good buffering capacity. Small differences in HLR between phases 1 and 2 did not alter wetland performance. This was demonstrated by a lack of change in wetland outflow quality for the control cells. The improved quality in outflow wastewater of planted cells may be due to plant uptake.

Phenol removal efficiency was higher when macrophytes were in an exponential (active) growth stage compared to the steady growth phase for all species (Fig. 6.9). During the former stage, plant uptake rate was significant with Phragmites, papyrus and Typha taking up 5.4 mg/day, 7.2 mg/day, and 12.7 mg/day respectively. These accounted for about 10 %, 13 % and 23 % respectively of the phenol removal. In the later growth stage, however, uptake was minimal ranging from 0.2 mg/day (papyrus) to 0.8 mg/day (Phragmites) accounting for only 0.3 % to 1 % of the phenol removed. Hence in the steady plant growth stage there was no significant difference between planted and unplanted wetland cells. This demonstrates that the microbial communities involved in both cases were similar and capable of pollutant degradation.

Phenol removal processes and reaction rates

This study reports for the first time, the contribution of various processes responsible for phenol removal and reaction rates for phenol reduction in constructed wetlands receiving pulp and paper mill wastewater. Microbial degradation was the major process for the removal of phenol in the constructed wetland. This was illustrated by the difference in removal efficiency between gravel buckets inactivated with formalin and those that were biologically active. Various microorganisms are responsible for the degradation of phenolic substances. Although the microorganisms were not identified in this study, various studies with pure and mixed cultures report the role of fungal and bacterial species in phenol degradation (Fountoulakis *et al.*, 2002; Cespedes *et al.*, 1996; Lee *et al.*, 1994). Whilst the microbial degradation was estimated at 60 %, ultimate biodegradation in a mature wetland may be higher as some of the sedimented and/or adsorbed phenols may be re-cycled and microbially degraded. Buckets with only papyrus had similar phenol removal efficiency as those that had papyrus and gravel. This observation could have beneficial implications for cost considerations when constructing a full-scale wetland.

Sedimentation and/or adsorption were found to play a significant role in phenol removal (about 30 % at 3-day HRT). Adsorption of organic compounds, including phenols may occur via chemisorptions or physical adsorption (DeBusk, 1999; Calace *et al.*, 2002). Plant uptake in buckets with papyrus was significant accounting for 10 % at 3-day HRT. However this role is only maintained during active plant growth. Similar observations were made in the wetland cells. Volatilisation accounted for up to 45 % phenol removal in buckets having wastewater only. This is higher than the 30 % reported by Polprasert *et al.* (1996). The difference may be attributed to the container size and small volume (20 litres), and lower concentrations of phenol in the effluent used in this study. The mechanism of phenol loss was however similar; there appeared to be no evaporative concentration and phenol seemed to be lost with the bulk water (evaporative loss). In the FWS cells plant shading may have limited phenol volatilisation.

Phenol removal in the wetland followed first order model (Fig. 6.7). Using the exponential curve-fitted equation in the figure, for an outflow phenol concentration of 0.05 mg/l or less (discharge regulations) the study wetland should have been at least 4 m in length. The first order areal removal rate, k, for phenol reduction was on average 27 m/yr (0.07 m/day) for planted SSF cells (series B-D). Series E cells had a rate constant of 38 m/yr (FWS) and 44 m/yr (SSF). The volumetric removal rate, k_v, for batch operation was on average 0.57 d^{-1} for *Phragmites* and papyrus cells. There are no removal rates reported in literature for phenol in constructed wetlands. However the k_v value was higher than the range (0.08 – 0.39 day^{-1}) reported by D'Angelo and Reddy (2000) in their laboratory experiments for transformation of pentachlorophenol using wetlands soils under various environmental conditions.

The measured data for phenol concentration in the wetland outflow (Figure 6.8) did not follow inflow peaks indicating that the constructed wetland offered consistent buffering even at the highest inflow concentration of 1.3 mg/l encountered in this study. The simulated wetland phenol outflow on the other hand did not match the measured data perfectly. The first order model assumes steady state and uses influent flow and concentrations coupled with ideal or non-ideal plug flow to predict exponential pollutant behaviour (Rousseau, 2005). This, according to Rousseau (2005) and Kadlec (1997, 2000), is a major drawback of this type of model since small-scale wetlands are subject to large influent variations leading to non-steady state conditions. In our study gate valves controlled the wetland wastewater inflow rate but inflow phenol concentrations were highly variable and depended on the on-going operations in the factory and the performance of the PANPAPER Mills wastewater ponds. Furthermore, variations in precipitation and evapotranspiration were evident on a daily and/or weekly basis. Kadlec (1997) cautions that the model uses long-term averages of variables such as flow rate and predictions therefore tend to be direct responses to inlet and outlet peaks and valleys. Potential uptake by plants is also not considered in the model.

Implication of wetland performance for the Nzoia river water quality

The measured water quality of Nzoia River with respect to phenol concentrations (Chapter 2) indicated an increase of 100 % downstream (3 km) of PANPAPER Mills's effluent discharge point compared to an upstream sampling point (500 m). If a constructed wetland of similar performance to the one in this study were incorporated (as a tertiary treatment stage) to purify the wastewater before discharge into the river, phenol concentrations in the final effluent would be reduced to an average of 0.07 mg/l and 0.053 mg/l for continuous flow (*C. papyrus*, active plant growth stage) and batch loading (*Phragmites*, 5-day HRT) respectively. Figure 6.10 illustrates the phenol concentrations along the existing wastewater treatment ponds plus a constructed wetland. Subsequently, the downstream phenol concentrations in the river would decrease by nearly 50%. The downstream and upstream would hence be similar (differing by only 7 – 9%) indicating that there would be much less pollution compared to the current status. The constructed wetland is therefore suitable for buffering the Nzoia River from water quality degradation with respect to phenols. Figure 6.11 gives the predicted concentrations for the river. A comparison is made with a "business-as-usual" scenario using the current PANPAPER Mills treatment ponds only.

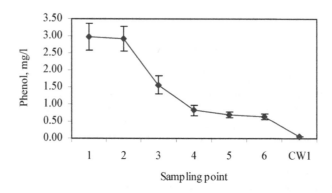

Fig. 6.10 Expected spatial variations in outflow phenol concentration at various stages of PANPAPER Mills effluent treatment system (Chapter 2) if a constructed wetland is incorporated. CW1 = a constructed wetland that is planted with *Phragmites* and is batch 5-day cycle operated.

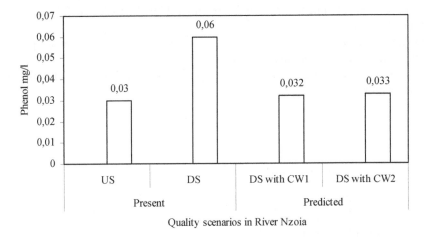

Legend:
US   = upstream sampling point in the river
DS   = downstream sampling point
CW1 = constructed wetland with batch loading and *Phragmites*
CW2 = constructed wetland with continuous flow and *C. papyrus*

Fig. 6.11 Predicted phenol concentrations downstream of effluent discharge point if treatment were to incorporate a constructed wetland with the present PANPAPER Mills ponds. A comparison is made with a "business-as-usual" scenario (using the existing ponds only) for low flow in the river (10 m³/s).

Implications for design, operation and maintenance

The reaction rate constant and other data from this study were used to derive preliminary design parameters for a full-scale SSF constructed wetland based on phenol removal alone. Using Equation 6.2 and substituting for hydraulic loading rate (HLR) and rearranging the equation to:

$$A = \frac{0.0365Q}{k} \times \ln C_i/C_e$$, the required area was determined to be 106.5 ha for

a full-scale wetland to polish PANPAPER Mills' wastewater. The corresponding HLR is 3.5 cm/day. However, increasing the HLR to say 9.8 cm/day (ca 35.8 m/yr) as was used in this study can minimize the area. The resulting area would then be approximately 35 ha. Wetland efficiency in removing phenols was raised by 10 – 20 % during periods of active plant growth. Maintaining plants in an active growth stage by sequential harvesting of shoots should exploit this role.

Conclusion

From the findings of our study on phenol dynamics in a constructed wetland we conclude as follows:
1. The performance of the constructed wetland indicated good buffering even during the highest inflow phenol concentration of 1.3 mg/l encountered during the study and the highest hydraulic loading rate of 9.8 cm/day in the wetland. For batch operation, optimal removal was achieved at 5-day hydraulic retention time.
2. We predict that by integrating a full-scale constructed wetland of similar performance as the one in this study into the system the current state of pollution in the Nzoia River would be reduced by more than 90 %.

3. The major processes of phenol removal are microbial breakdown followed by sedimentation/adsorption.
4. Phenol removal is enhanced by wetland age and presence of aquatic macrophytes especially when they are at an exponential growth stage.
5. There was no significant difference in the overall performance of wetland cells planted with different species.

References

Abira M. A, Ngirigacha H. W. and van Bruggen J.J.A. 2003. Preliminary investigation of the potential of four tropical emergent macrophytes for treatment of pre-treated pulp and papermill wastewater in Kenya. *Water Science and Technology* 48 (5): 223 - 231.

Abira, M.A., van Bruggen, J.J.A. and Denny, P. 2005. Potential of a tropical subsurface constructed wetland to remove phenol from pre-treated pulp and papermill wastewater. *Water Science and Technology* 51 (9): 173 - 176.

APHA, 1995. *Standard methods for the analysis of water and wastewater*, 19th edition. American Public Health Association, Washington DC.

Banerjee, G. 1997. Treatment of phenolic wastewater in RBC reactor. *Water Research* 31 (4): 705 - 715.

Calace, N., Nardi, E., Petronio, B.O. and Pietroletti, M. 2002. Adsorption of phenols by paper mill sludges. *Environmental Pollution* 118 (3): 315 – 319.

Cespedes, R., Maturana, A., Buman, U., Bronfman, M. and Gonzalez, B. 1996. Microbial removal of chlorinated phenols during aerobic treatment of effluents from pine Kraft pulps bleached with chlorine-based chemical, with or without hemicellulases. *Applied Microbiology and Biotechnology* 46: 631 – 637.

D'Angelo, E.M. and Reddy, K.R. 2000. Aerobic and anaerobic transformation of pentachlorophenol in wetland soils. *Soil Science Society of America Journal* 64: 933 – 943.

DeBusk, W.F., 1999. Wastewater treatment wetlands: applications and treatment efficiency. *Fact sheet SL156*, Soil and Water Science Department, Institute of Food and Agricultural Sciences, University of Florida. Website http://edis.ifas.ufl.edu

Eckenfelder, W. W. Jr. 2000. *Industrial Water Pollution Control*, 3rd edition, McGraw-Hill, New York U.S.A.

Eckhardt, D.A.V., Surface, J.M. and Peverly, J.H. 1999. A constructed wetland system for treatment of landfill leachate, Monroe County, New York. In: Mulamoottil, G., McBean, E.A and Rovers, F. (eds.): *Constructed wetlands for the treatment of landfill leachates*, Lewis publishers, Boca Raton, pp. 205 – 222.

Frigon, J.C., Cimpoia, R. and Guiot, S.R. 2003. Sequential anaerobic/aerobic bio-treatment of bark leachate. *Water Science and Technology* 48 (6): 203 – 209.

Fountoulakis, M.S., Dokianakis, S.N., Kornaros, M.E., Aggelis, G.G. and Lyberatos, G. 2002. Removal of phenolics in olive mill wastewaters using the white-rot fungus *Pleurotus ostreatus*. *Water Research* 36: 4735 - 4744.

Kadlec, R.H. and Knight, R.L. 1996. *Treatment wetlands*. CRC Press Inc., Boca Raton, Florida, 893 pp.

Kadlec, R.H. 1997. Deterministic and stochastic aspects of constructed wetland performance and design. *Water Science and Technology* 35 (5): 149 – 156.

Kadlec, R.H. 2000. The inadequacy of first order treatment wetland models. *Ecological Engineering* 15: 105 - 119.

Lee, C.M., Lu, C.J. and Chuang, M.S. 1994. Effects of immobilized cells on the biodegradation of chlorinated phenols. *Water Science and Technology* 30 (9): 87 - 90.

Masbough, A., Frankowski, K., Hall, K.J. and Duff, S.J.B., 2005. The effectiveness of constructed wetland for treatment of wood waste leachate. *Ecological Engineering* 25: 552 – 566.

Mogere, S. N. 2000. Assessment of chironomus (Diptera: chironomidae) larval head deformities as possible bioindicators for pollutants in sediments of R. Nzoia, Kenya. M.Phil. Thesis, Moi University, Eldoret, Kenya.

Polprasert, C., Dan, N.P. and Thayalakumaran, N. 1996. Application of Constructed wetlands to treat some toxic wastewaters under tropical conditions. *Water Science and Technology* 34 (11): 165 - 171.

Roberts, M. M. 2004. The influence of nutrient enrichment on the purification of pulp and paper mill wastewater by various plant species in a tropical constructed wetland. M.Sc. Thesis (SE 04.09), UNESCO-IHE, Delft, The Netherlands.

Rousseau, D., 2005. Performance of constructed treatment wetlands: model-based evaluation and impact of operation and maintenance. PhD Thesis, Ghent University, Ghent, Belgium, 300 pp.

Thut, R.N. 1993. Feasibility of treating pulp mill effluent with a constructed wetland. In: Moshiri, G.A. (ed.). *Constructed wetlands for water quality improvement*, CRC Press, Lewis Publishers, Boca Raton, pp 441 - 447.

US EPA, 1993. United States Environmental Protection Agency. Subsurface flow constructed wetlands for wastewater treatment: A technology assessment, EPA 832-R-93-008.

Chapter 7

Discussion, conclusions and outlook

Discussion, conclusions and outlook

Introduction

Natural and artificial wetland systems have been used as a cost-effective alternative to conventional wastewater treatment methods for improving final effluent quality (Hammer and Bastian, 1989; Kadlec and Knight, 1996; Kadlec *et al.*, 2000). Besides domestic application, the use of constructed wetlands has spread to many other fields including industrial effluent treatment, acid mine drainage, agricultural effluent, landfill leachate and road runoff. However, data and information on pulp and paper mill wastewater treatment in constructed wetlands are few while performance data that can guide design and operation under tropical environment conditions are lacking. It was hypothesised in this study: 1) that emergent aquatic macrophytes are suitable for treatment of the wastewater; 2) that hydrological factors influence the performance of the wetland system and 3) that the study constructed wetland would yield effluent of a quality that would comply with Kenyan discharge regulations. This study was undertaken to establish the efficacy and optimal performance data for a constructed wetland planted with local emergent macrophytes in treating pulp and paper mill wastewater under tropical environment conditions. The impact on the receiving Nzoia River, of integrating such a wetland with the present pond treatment system at the Pan African Paper Mills (E.A.) Limited, Webuye, Kenya (PANPAPER) was predicted from the latter's performance data and the measured quality of the river water prior to and after discharge of the treated effluent.

PANPAPER Mills ponds

The process wastewater emanating from PANPAPER Mills is laden with pollutants such as organic matter, suspended solids, and phenols. The raw effluent exhibits high variation in quality due to differences in the main processes taking place at the mill. Pollutant concentrations are particularly high during desludging (maintenance dredging). Wastewater treatment ponds of PANPAPER Mills are performing well as per its type and design. However, the final effluent discharged into the Nzoia River, despite being weaker than the raw wastewater does not comply with the national discharge limits.

Aerated lagoon or pond systems such as the ones at PANPAPER Mills is one of the options for conventional treatment of pulp and paper mill wastewater and have to a large extent become the industry standard for wastewater treatment (Johnston *et al.*, 1996). The performance of the PANPAPER Mills lagoons with respect to BOD, COD and phenols (Chapter 2) was comparable to that reported for similar ones by Stuthridge *et al.* (1991). Despite their performance the lagoons seem inadequate for ensuring compliance with discharge regulations in Kenya leading to poor water quality even more than 3 km downstream in the recipient river. In the United States of America, pulp mill wastewater quality improvement has been achieved through additional treatment in constructed wetlands in order to meet stringent guidelines and protect the recipient streams especially during low flow (Thut, 1989.) A similar buffer is necessary at PANPAPER Mills to protect the Nzoia River.

In this chapter I give an appraisal of the main findings on the performance of trial constructed wetlands and the implication for the water quality of River Nzoia, of discharging wastewater from it being integrated (as a tertiary treatment stage) with the existing PANPAPER Mills wastewater ponds. Design criteria and appropriate operation and maintenance conditions for a full-scale facility are recommended. Finally, with some optimism, the outlook on opportunities for the use of and research into constructed wetlands is presented.

Purification efficiency

The constructed wetland system exhibited good performance (removal efficiency) with respect to all major pollutants monitored viz., organic matter and suspended solids (Chapter 5), phenols (Chapter 6), and nitrogen and phosphorus (Chapter 4). The wetland effluent quality met the prescribed regulations (Water Resources Management Rules, 2007) except for COD. The system showed high buffering capacity during peak inflows (shock loads) such as shown in Fig 6.8 (Chapter 6) especially during maintenance dredging. Notwithstanding, the outflow pollutant concentrations on such occasions exceeded discharge guidelines. Such a situation can be mitigated, in a full-scale facility, by creating a deep-water zone at the inlet end (Kadlec *et al.*, 2000) to settle suspended solids. I would also recommend another deep-water zone at the tail end (outflow) of the wetland where additional settling could remove more suspended solids and the associated COD attributed to the presence of cellulose fibres, lignins and lignin degradation products (Edde, 1984).

The nature of the COD in the wetland is continually changing. Microbial consumption of readily and slowly biodegradable COD eventually leaves residual inert and hard to degrade COD (Edde, 1984). Such are degradable by fungi especially the white-rot fungi (Kondo, 1998; Fountoulakis *et al.*, 2002). According to Kondo (1998) the fungal populations in the wetland may be increased by inoculation with wood mulch compost. Wetlands inoculated with wood mulch compost show improved COD degradation (e.g. Hammer *et al.*, 1993). Perhaps such inoculation could be done at the beginning as the wetland is set up. With long-term operation, accumulation and/or decomposition of litter will increase fungal populations and as such may enhance dissolved COD removal. Sustained removal of residual dissolved COD may also be achieved by using white wood ash (Tantemsapya *et al.*, 2004). White wood ash is generated as a solid waste from the bark (multi-fuel) boiler at the pulp mill and would therefore not have extra costs for operating the wetland system. It could be packed in sacks and placed at the outflow of the constructed wetland before the extra settling lagoon (See recommendation below). Some studies such as that by Oeller *et al.* (1997) have recommended the use of ozone or combination of ozone and ultraviolet light treatment to increase COD degradation. This is not only inappropriate especially for a developing country like Kenya with inadequate energy supply but also very costly and needs highly skilled labour.

The constructed wetland in this study acted as a tertiary system receiving the final treated wastewater from PANPAPER Mills's stabilisation pond. I intended, as part of the study to operate some of the wetland cells (with *Typha domingensis*) as a secondary system as well, in order to establish how far pre-treatment is necessary for satisfactory treatment in the wetland. However, an unforeseen temporary factory shut down in wood pulping upset the experiment. In reported studies on wetlands used for pulp and paper mill wastewater treatment only one system undertook secondary treatment (Boyd *et al.*, 1993) and data on wetlands performance of such systems are still lacking. Nevertheless, the buffering effect of this study's wetland during peak loadings is a pointer to possible effective performance in a secondary system.

The wetland's performance with respect to phenol removal was better when the system age was more than two years compared to that in the first year (Chapter 6; Abira *et al.*, 2005). The difference is attributed to the more extensive spread of macrophyte roots and the time needed for establishment of a microbial environment conducive for bio-degradation of phenol in the wetland.

What then would be the expected long-term performance of the constructed wetland system? *Typha* cells had consistently higher TSS removal efficiency than *Phragmites* and papyrus (Chapter 5). This difference may be attributed to differences in plant root penetration (Chapter 4). By the end of this study when the wetland was three years of age only the *Typha* roots had penetrated the entire bed depth (approx. 30 cm). It may be postulated that the purification efficiency would improve with age and maturity of the wetland: with increased plant root zone aeration (Brix, 1997) adsorption and uptake of phenols would be enhanced.

However, continuous maintenance (see below) is necessary to sustain high purification efficiency as well as to prolong the life of the wetland system (Rousseau, 2005).

Processes and influencing factors

In constructed wetlands the reduction of organic matter and suspended solids is accomplished by diverse treatment mechanisms namely: sedimentation, filtration, adsorption, microbial interactions, and uptake by vegetation (Watson *et al.*, 1989; Cooper *et al.*, 1996; Kadlec *et al.*, 2000). Besides the inflow and outflow, the solids mass balance is mainly influenced by filtration, plant and microbial growth and decay as well as the prevailing hydrological and hydraulic conditions. Dissolved organic matter is removed mainly by microbial degradation and to a lesser extent, by uptake from plants. Microbial degradation of COD (based on the maximum dissolved COD removed) was low (up to 42 %) whereas degradation of phenols was high (up to 70%).

Nitrogen removal in temperate-climate treatment wetlands is mainly by sequential nitrification-denitrification (Kadlec and Knight, 1996; Reddy *et al.*, 1989) with adsorption and plant uptake playing minor roles. In low loaded systems (e.g. Toet *et al.*, 2005) and those in tropical climatic regions (e.g. Okurut, 2001; Koottatep and Polprasert, 1997; Tanner, 1996) where plant growth is favoured by all-year-round warm temperatures, plant uptake plays a significant role (see discussion below). Sedimentation/adsorption and plant uptake were the dominant processes in phosphorus removal.

Influence of system operation mode, hydraulic loading rate/retention time
The load-drain batch mode of operation was initially adopted with the assumption that it would aerate the wetland bed and enhance organic matter degradation (e.g. Stein *et al.*, 2003). Although some aeration occurred as indicated by higher redox potentials during batch operation (Chapter 5), the dissolved oxygen concentration remained low in the wetland resulting in a strong anaerobic and reducing environment. The control (unplanted) cells exhibited lower redox potential than planted cells an indication that plants contributed to oxygenation of the wetland beds (Brix, 1997). It is assumed that anoxic and anaerobic degradation processes were predominant (Kadlec *et al.*, 2000).

COD and TSS removal efficiency were significantly higher in the continuous flow compared to the batch-loading mode of operation. The lower efficiencies may be attributed to resuspension of previously deposited solids in the gravel voids (Kadlec and Knight, 1996). BOD removal, however, was not significantly different between the two modes of operation. Overall, the efficacy of the constructed wetland in treating the wastewater was higher during continuous flow operation. In a full-scale system topping up to compensate for evapotranspiration losses (ET) would not normally be undertaken. The ET losses during each batch cycle would therefore be excessively high. This coupled with the poorer performance in organic matter and suspended solids removal makes batch operation unattractive. I therefore recommend continuous flow operation for a full-scale wetland.

The wetland displayed good buffering capacity with respect to organic matter, TSS, phenols and nutrients at the higher (ca double) hydraulic loading rate as indicated by similarity in outflow concentrations between series E and those with series A-D (Chapters 4 – 6).

Influence of plants
Plants are an integral part of wetlands. Their ability to purify wastewater depends on the ability to grow rapidly and to form extensive root mats, which take up water, nutrients and other substances while providing vital oxygen in the root zone, and attachment surface for microorganisms. In the study wetland pollutant removal was significantly higher in planted than unplanted cells for phenols, nutrients and total suspended solids, especially when plants were in the exponential growth stage.

It is generally believed that nutrient removal via plant uptake is only important in lightly loaded systems such as the one in this study (Toet *et al.*, 2005; Vymazal, 2001; Kadlec, 2004, Kadlec *et al.*, 2000) and that plant uptake constitutes temporary storage with subsequent release at senescence unless the vegetation is harvested at the stationary growth phase (Brix, 1997). For constructed wetlands studied in the temperate climatic conditions this harvesting is thought to be detrimental to plant resilience (e.g. Geller, 1997; Vymazal, 2001). In contrast, constructed wetlands in the tropics are favoured by all-year-round plant growth and hence the contribution of plant uptake to the overall nutrient removal is significant even in high loaded systems (Okurut, 2001; Koottatep and Polprasert, 1997). In the study by Okurut (2001) plant growth and nutrient uptake in presettled sewage was sustained even with shoot harvests up to four times a year. In this study up to 25 % of nitrogen and phosphorus removal efficiency may be attributed to plant uptake with subsequent harvesting of above-ground biomass.

The significance of plants was also evident in the removal of TSS. *Typha domingensis* plants whose roots penetrated the entire wetland bed had consistently higher TSS removal efficiency compared to *Phragmites mauritianus* and *Cyperus papyrus*. The role of plants in pollutant removal depends on their growth rate. *Typha* and *Phragmites* achieved adequate growth during the study period (Chapter 4). In addition pollutant removal efficiency was higher in cells planted with these macrophytes than in papyrus. However, papyrus growing in wastewater buckets without gravel was found to remove phenol as efficiently as those in gravel buckets. In the experimental buckets without gravel papyrus had extensive root growth. Use of larger size gravel (ca 12.5 mm) or a mixture of large and small (e.g. 6.25 mm and 12.5 mm in a 1:1 ratio) may enable faster penetration of the gravel media in rooted *C. papyrus*.

Implications

Water quality of Nzoia River
The concentration of potential pollutants in the final effluent of PANPAPER Mills exceeds that allowable by the regulating authority in Kenya. In a "business-as-usual" scenario the discharged final effluent causes an increase in the concentration of pollutants downstream (3 km) compared to an upstream location (Chapters 3 – 6). The increase is highest during low flow in the River, usually in the dry season (January to March). This was reflected in an increase in the concentration of various pollutants by between 20 and 120 % at the downstream sampling point during the month of March. From the findings of this study I predict that the quality of water in River Nzoia downstream of the discharge point would improve significantly if a constructed wetland with similar performance as that in the pilot study were to be integrated, as a tertiary stage, with the existing treatment ponds. Such an intervention would decrease pollution by more than 90 % for phenols. The new downstream and upstream concentrations would be similar (Fig. 6.11, Chapter 6). Furthermore, the downstream concentrations of TSS, COD and BOD would be reduced by 44 %, 30 % and 63 %, respectively.

The discharge of nitrogen and phosphorus is now regulated in Kenya with guidelines of 10 mg/l and 2 mg/l respectively (Water Resources Management Rules, 2007). The nutrient concentrations in the treated effluent of PANPAPER Mills are low. However, due to the large quantity of effluent discharged there is a potential for enrichment especially in impoundments downstream, including Lake Victoria. With the wetland intervention proposed in this study nitrogen concentration downstream of effluent discharge would remain the same as that in the upstream location (100 % pollution reduction). Predicted phosphorus concentrations in the river downstream would reduce by 50 % of the current level.

Scaling-up: design, construction, operation and maintenance recommendations
Due to the lower removal efficiency and poorer plant growth during the batch operation mode as compared to continuous flow I recommend the adoption of continuous flow for a full-scale constructed wetland. In the sizing of a treatment wetland with steady flow the appropriate temperature and rate equations (Kadlec and Knight, 1996; Kadlec *et al*; 2000) are used in

combination with the water mass balance. This study has reported rate constants for phenol and BOD removal as well as the non-zero background concentrations for BOD in a horizontal subsurface flow constructed wetland (Chapters 5 & 6). For phenols the background concentration is assumed to be zero since occasionally phenols were not detectable in the wetland outflow. In this study even at higher loading rates 9.3 – 9.8 cm/day for example as with constructed wetland cells Series E the wetland outflow concentrations for phenols, nutrients, BOD, COD and TSS were comparable to those with lower loading rates (4.1 - 4.9 cm/day). I therefore recommend a hydraulic loading rate of about 10 cm/day for the full-scale wetland.

The BOD reaction rate constant obtained at the higher loading (0.114 m/day or 42 m/yr) is higher than that obtained for phenols (ca 36 m/yr) in the same wetland cells (Chapters 5&6). Most wetlands' design is based on BOD removal. However, Kadlec and Knight (1996) recommend wetland sizing on the regulated parameter requiring the largest area (with a lower reaction rate constant). In the case of wastewater from PANPAPER Mills the size recommended for phenol removal in Chapter 6 (approx. 35 ha) would be appropriate for BOD removal as well.

Incorporating the water balance (Kadlec, 1997; Kadlec et al., 2000) to cater for evaporative losses gives the average of the wetland inflow and outflow wastewater as approximately 29,000 m³/day. The required land area is thus reduced to 29 ha. Based on the performance of the study wetland and that of the PANPAPER Mills stabilisation ponds (3 and 4), I recommend that the latter be replaced with a constructed wetland. The total area occupied by the two ponds is 15.5 ha (Chapter 2). Therefore, the additional land area required for the constructed wetland is 13.5 ha. A summary of the guidelines for the design of the full-scale wetland is given in Table 7.1

Table 7.1 Preliminary design parameters, based on phenol removal, for a full-scale SSF wetland to treat PANPAPER Mills wastewater. Assumption: Plug flow and zero-background phenol concentration.

Parameter	Mean	Remarks
Design flow (Q), m³/day	34, 201	Q = 29,000 m³/day when ET losses in the wetland (this study) are considered.
Influent concentration, mg/l	0.50	4-year mean at PANPAPER Mills (this study)
Target effluent concentration, mg/l	0.05	Kenyan regulations (Water Resources Management Rules, 2007)
Reduction fraction to target or mean removal efficiency	0.9	This study
Areal rate constant, k, m/yr	36	This study
Required area (approximate), ha	35	Calculated
	(29)	(Allowing for ET losses - this study)
	(13.5)	(If PANPAPER ponds 3 & 4 are replaced with constructed wetland)
Hydraulic loading rate, q, cm/day	10	This study
Water depth (approximate), cm	30	This study. (May be increased to 50 cm for floating papyrus)
Water Temperature, °C	24	This study

Guidelines for construction and set up of constructed wetlands are given in various publications (e.g. WRc, 1990; EPA, 1993; Crites, 1994; Kadlec and knight, 1996; Kadlec *et al.*, 2000) and are hence not discussed here. However, due to the high variation in the quality of wastewater from PANPAPER Mills's treatment ponds and their periodic maintenance dredging it is prudent to have a deep-water zone at the wetland inlet that can collect large amounts of sediments and facilitate clean out (Kadlec *et al.*, 2000). I therefore recommend that a free water surface flow cell with a deep inlet zone should precede the subsurface flow wetland cells (Fig 7.1).

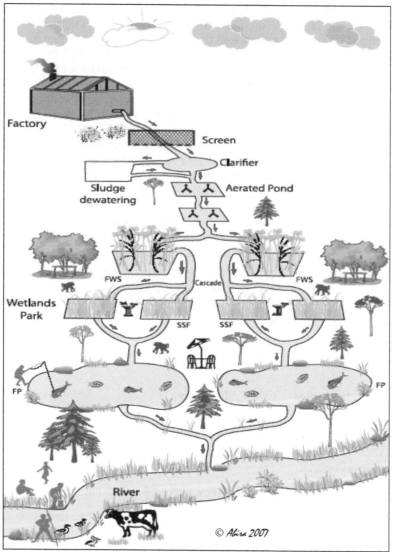

Fig 7.1 Conceptual diagram of a full-scale constructed wetland integrated with stabilisation ponds to treat pulp and paper mill wastewater. FP = Fish Pond, FWS = Free Water Surface flow wetland, SSF = Sub Surface Flow wetland.

As discussed above, *Cyperus papyrus* growth in the constructed wetland would be enhanced in a non-gravel system. It may possibly be used in the free water surface flow cell. Besides local availability wetland macrophytes are selected based on their growth rate and product use. All three macrophyte species used in this study find a variety of uses among the local communities.

In addition PANPAPER Mills could use them as a supplementary energy source for fuelling boilers depending on their calorific value. PANPAPER Mills has established that papyrus culms have a good calorific value (Personal communication with PANPAPER Mills' quality control manager). The drawback, however, is the product bulkiness and transportation costs from far-sited natural wetlands. The close proximity of a constructed wetland would eliminate this cost. It is probable that *Phragmites mauritianus* would also provide a good energy source. *Typha domingensis* is non-woody and may not be a suitable fuel source.

The efficiency of phenol and nutrient removal in this study was higher when plants were at an exponential growth stage. Regular harvesting of plants shoots would therefore maintain this role. However, harvest interval is crucial so as not to compromise plant regeneration and vitality. There is need to carry out studies to establish a suitable harvest interval due to the low nitrogen in the wastewater. At the moment I would recommend harvesting at 12-monthly intervals in rotation so that at least two thirds of the treatment bed is actively growing.

Constructed wetlands have low maintenance requirements. However, Rousseau (2005) cautions that they are not a "build-and-forget" technology and continuous maintenance is necessary in order to sustain high purification efficiency as well as to prolong the life of the wetland system. It is necessary to desludge the inlet zones; remove litter and harvest plant shoots on time.

Conclusions

From the main findings of this study and the foregoing discussion I can conclude as follows:

- ➢ The performance of the constructed wetland indicated good buffering and effectively removed phenols, BOD, TSS, nitrogen and phosphorus from pre-treated pulp and paper mill wastewater to concentrations below that prescribed by the regulating authority in Kenya. However, COD removal was unsatisfactory and the new guideline of 100 mg/l for COD was not achieved.

- ➢ Phenol and nutrient removal efficiencies were enhanced by the presence of aquatic macrophytes especially when they are at an exponential growth stage.

- ➢ The major processes of phenol removal are microbial breakdown followed by sedimentation/adsorption and plant uptake.

- ➢ Integrating a full-scale constructed wetland with the current treatment system of PANPAPER Mills significantly improves the effluent quality and would reduce pollution in the Nzoia river from current levels by 90%, 30 %, 44 %, 63 %, 100 % and 50 % for phenols, COD, TSS, BOD, nitrogen and phosphorus, respectively.

- ➢ Evapotranspiration (ET) is an important component of outputs in the water budget of the wetland system. It should therefore be an integral part in wetland design in the tropics.

- ➢ ET rates are different for different aquatic plant species. Therefore ET may be an important selection criterion for plant species to be used in a constructed wetland used for wastewater treatment especially in drier climates.

- ➢ Plant tissue nutrient concentrations were lower than in healthy natural wetland plants. Nevertheless, *Typha* and *Phragmites* had satisfactory biomass production. The growth of papyrus was sub-optimal. *Phragmites, Typha* and papyrus (in that order) are all suitable species for the effective removal of nutrients. A mixture of all three species (or whichever is available in the locality) might provide the best long-term option.

➤ Cycling of nutrients in sediments and/or in senescing/decaying plant organs are an important source for sustaining plant growth in a low loaded system such as the one in this study. Therefore harvesting of shoots should be appropriately timed to avoid depletion of the nutrient pool.

➤ The required land area for a full-scale constructed wetland is 29 ha. However if PANPAPER ponds 3 and 4 are converted to wetlands, the additional land area required is only 13.5 ha.

Outlook

Opportunities for the use of constructed wetlands
This study was undertaken to establish the efficacy of the constructed wetland technology in purifying pulp and paper mill with the goal of improving the management of industrial effluents in Kenya, and in the east African region. The findings of this study and the predicted improvement of water quality in the Nzoia River downstream of PANPAPER Mills discharge if the technology is adopted is indeed an ace for the management of wastewater in Kenya and in the region.

Constructed wetlands are generally less expensive to build compared to other energy intensive wastewater treatment technology (Kadlec *et al.*, 2000). The major investment cost is land but the long lifespan of the wetland would offset this. PANPAPER Mills already has adequate land at its present site. The investment cost of the constructed wetland including excavation and compacting works, liner, plants, plumbing and gravel but excluding the cost of land is estimated[*] at about US$ 2.1 million. The benefits of compliance with wastewater discharge guidelines certainly include improved water quality for ecosystem integrity and public health on the one hand and a big image boost for the factory. Other benefits are by product use, creation of wildlife habitat and recreation.

The findings of this study coupled with the positive findings of other studies in east Africa on the suitability of wetlands for treatment of domestic wastewater in the tropics (e.g. Kipkemboi *et al.*, 2002; Okurut, 2000; Mashauri *et al.*, 2000; Nyakang'o and van Bruggen, 1999; Kansiime and Nalubega, 1999) is likely to stimulate further the adoption of wetland technology. As suggested by Denny (1997) constructed wetlands could, therefore be the technology of choice in wastewater treatment to protect both surface and groundwater resources in various situations including the following:

➤ Municipal treatment facilities - for secondary or tertiary treatment,
➤ Industrial establishments as secondary treatment or as tertiary polishing stage,
➤ Institutions such as hospitals, hotels, restaurants, schools and colleges that are remotely located from sewerage facilities,
➤ Individual households with onsite sanitation,
➤ Catchment management e.g. agro-chemical buffer zones in farms bordering important streams/environmentally sensitive lakes.

---

[*] Rough estimate not based on site characteristics and other requirements such as vegetation clearing, drainage, fencing etc.

Research opportunities

This study was not without limitations. It was carried out in a pilot constructed wetland that was designed based on the rules of thumb and BOD as the primary pollutant. In Chapter 6 it was found that the length of the subsurface cells should have been at least 4 m in order to remove phenols completely. Appropriate wetland design includes provision for adequate hydraulic sizing. The tracer study was not conclusive in elucidating the hydraulic characteristics. In terms of pollutant removal efficiency, COD's removal was unsatisfactory. The strategy adopted for the research study where all pollutants were concurrently investigated did not offer flexibility for optimising the removal of all. There is need to establish the conditions necessary for better COD removal with the methods suggested above. Phenol removal is partly via adsorption. Although simple phenols are degraded completely it is necessary to establish the fate of other more complex ones and potential for their release by the sediment into the effluent stream. The fate of phenols taken up by the plants: whether volatilised or stored with a potential for later release needs to be established. Nutrient dynamics in the wetland should be studied including mineralisation rates and the periodicity for biomass retrieval.

Finally monitoring of pollutant flows in various wetland compartments not limited to in-out sampling would generate data and information that will serve as a basis for optimising model-based approaches to the design and operation of constructed wetlands for treatment of pulp and paper mill wastewater in the tropics.

References

Abira, M.A., van Bruggen, J.J.A. and Denny, P. 2005. Potential of a tropical subsurface constructed wetland to remove phenol from pre-treated pulp and papermill wastewater. *Water Science and Technology* 51 (9): 173 - 176.

Boyd, J., McDonald, O., Hatcher, D., Portier, R.J., and Conway, R.B. 1993. Interfacing constructed wetlands with traditional wastewater biotreatment systems. In: Moshiri, G.A. (ed.). *Constructed wetlands for water quality improvement.* CRC Press, Lewis Publishers, Michigan, pp 453 – 460.

Brix, H. 1997. Do macrophytes play a role in constructed treatment wetlands? *Water Science and Technology* 35 (5): 11 - 17.

Cooper, P.F., Job, G.D., Green, M.B., and Shuttes, R.B.E. 1996. *Reed beds and constructed wetlands for wastewater treatment.* WRc plc. Swindon, Wiltshire, U.K., 184, pp.

Crites, R.W. 1994. Design criteria and practice for constructed wetlands. *Water Science and Technology* 29 (4): 1 - 6.

Denny, P. 1997. Implementation of constructed wetlands in developing countries. *Water Science and Technology*, 35 (5): 27 - 34

Edde, H. 1984. *Environmental control for pulp and paper mills.* Noyes Publications, Park ridge, New Jersey, USA. 500 pp.

EPA, 1993. *Subsurface flow constructed wetlands for wastewater treatment: A technology assessment.* U.S. Environmental Protection Agency EPA 832-R-93.

Fountoulakis, M.S., Dokianakis, S.N., Kornaros, M.E., Aggelis, G.G. and Lyberatos, G. 2002. Removal of phenolics in olive mill wastewaters using the white-rot fungus *Pleurotus ostreatus. Water Research* 36: 4735 - 4744.

Hammer, D. A. and Bastian R. K. 1989. Wetland Ecosystems: natural water purifiers? In: Hammer, D. A. (ed.), *Constructed wetlands for wastewater treatment: municipal, industrial and agricultural,* Lewis Michigan, U.S.A., pp 5 - 19.

Hammer, D.A., Pullin, B.P., Mc Murry, D.K. and Lee, J.W. 1993. Testing colour removal from pulp mill wastewaters with constructed wetlands. In: Moshiri, G.A. (ed.). *Constructed wetlands for water quality improvement.* CRC Press, Lewis Publishers, Michigan, pp 5 – 19.

Johnston, P.A., Stringer, R.L., Santillo, D., Stephenson, A.D., Labounskaia, I. Ph. & McCartney, H.M.A. 1996. *Towards zero-effluent pulp and paper production: the pivotal role of totally chlorine free bleaching.* Greenpeace Research Laboratories Technical Report 07/96. Publ: Greenpeace International, Amsterdam, ISBN 90-73361-32-X. 33pp.

Kadlec, R.H. 1997. Deterministic and stochastic aspects of constructed wetland performance and design. *Water Science and Technology* 35 (5): 149 – 156.

Kadlec, R.H. and Knight, R.L. 1996. *Treatment wetlands,* CRC Press Inc., Boca Raton, Florida, 893 pp.

Kadlec, R.H., Knight, R.L., Vymazal, J., Brix, H., Cooper, P. and Haberl, R. 2000. *Constructed wetlands for pollution control: processes, performance, design, and operation*. IWA specialist group on use of macrophytes in water pollution control. Scientific and technical report No.8, IWA publishing, London, UK, 156 pp.

Kansiime, F. and Nalubega, M. 1999. Wastewater treatment by a natural wetland: the Nakivubo swamp, Uganda. Processes and implications. Ph.D. thesis, A.A. Balkema Publishers, Rotterdam, The Netherlands, 300 pp.

Kipkemboi, J., Kansiime, F. and Denny, P. 2002. The response of *Cyperus papyrus* (L.) and *Miscanthidium violceum* (K. Schum.) Robyns to eutrophication in natural wetlands of Lake Victoria, Uganda. *African Journal of Aquatic Sciences* 27: 11 - 20.

Kondo, R. 1998. Waste treatment of Kraft effluent by white-rot fungi. In: Young, R. A. and Akhtar, M. (eds). *Environmentally friendly technologies for the pulp and paper industry*. John Wiley & Sons, Inc. pp. 515 – 539.

Koottatep, T. and Polprasert, C. 1997. Role of plant uptake on nitrogen removal in constructed wetlands located in the tropics. *Water Science and Technology* 37 (12): 1 - 8.

Mashauri, D.A., Mulungu, D.M.M. and Abdulhussin, B.S. 2000. Constructed wetland at the university of Dar es Salaam. *Water Science and Technology* 34 (4): 1135 - 1144.

Nyakang'o, J.B. and van Bruggen, J.J.A. 1999. Combination of a well functioning constructed wetland with a pleasing landscape design in Nairobi, Kenya. *Water Science and Technology* 40 (3): 249 - 256.

Oeller, H.J., Demel, I. and Weinberger, G. 1997. Reduction of COD in biologically treated paper mill effluents by means of combined ozone and ozone/UV reactor stages. *Water Science and Technology*, 35 (2-3): 269 - 276.

Okurut, T. O. 2000. A pilot study on municipal wastewater treatment using a constructed wetland in Uganda. PhD dissertation, Balkema publishers, Rotterdam, The Netherlands.

Okurut, T. O. 2001. Plant growth and nutrient uptake in a tropical constructed wetland. In: Vymazal, J. (Ed.). *Transformations of nutrients in natural and constructed wetlands*. Buckhuys Publishers, Leiden, The Netherlands, pp 451 - 462.

Reddy, K.R., Patrick Jr., W.H. and Lindau, C.W. 1989. Nitrification-denitrification at the plant-root sediment interface in wetlands. *Limnology Oceanography* 34: 1004 - 1013.

Rousseau, D., 2005. Performance of constructed treatment wetlands: model-based evaluation and impact of operation and maintenance. PhD Thesis, Ghent University, Ghent, Belgium, 300 pp.

Stein, O.R., Hook, P.B., Biederman, J.A., Allen, W.C. and Borden, D. J. 2003. Does batch operation enhance oxidation in subsurface constructed wetlands? *Water Science and Technology* 48 (5): 149 – 156.

Stuthridge, T.R., Campin, D.N., Langdon, A.G., Mackie, K.L., McFarlane, P.N. and Wilkins, A.L. 1991. Treatability of bleached Kraft pulp and paper mill wastewaters in a New Zealand aerated lagoon treatment system. *Water Science and Technology* 24 (3/4): 309 - 317.

Tanner, C. C. 1996. Plants for constructed wetland treatment systems-A comparison of the growth and nutrient uptake of eight emergent species. *Ecological Engineering* 7: 59 – 83.

Tantemsapya, N., Wirojanagud, W., and Sakolchai, S. 2004. Removal of colour, COD and lignin of pulp and paper wastewater using wood ash. *Songklanakarin Journal of Science and Technology*, 26 (Suppl.1): 1-12.

Thut, R.N. 1989. Utilisation of artificial marshes for the treatment of pulp mill effluents. In: Hammer, D.A. (ed), *Constructed wetlands for wastewater treatment: municipal, industrial, and agricultural*. Lewis, Chelsea, Michigan, U.S.A., pp 239 – 244.

Toet, S., Bouwman, M., Cevaal, A. and Verhoeven, J.T.A. 2005. Nutrient removal through autumn harvest of *Phragmites australis* and *Typha latifolia* shoots in relation to nutrient loading in a wetland system used for polishing sewage treatment plant effluent. *Journal of Environmental Science and Health part A* 40 (6-7): 1133 - 1156.

Vymazal, J. 2001. Types of constructed wetlands for wastewater treatment: their potential for nutrient removal. In: Vymazal, J. (ed.). *Transformations of nutrients in natural and constructed wetlands*. Buckhuys Publishers, Leiden, The Netherlands, pp 1 – 93.

Water Research Council (WRc). 1990. *European design and operations guidelines for reed bed systems*.

Watson, J.T., Reed, S.C., Kadlec, R.H., Knight, R.L., and Whitehouse, A.E. 1989. Performance expectations and loading rates for constructed wetlands. In: Hammer D.A. (ed.), *Constructed wetlands for wastewater treatment: municipal, industrial and agricultural*, Lewis Publishers, Chelsea, Michigan, U.S.A., pp. 319 - 351.

Water Resources Management Rules, 2007. Kenya Gazette Supplement No. 92, Legislative Supplement No. 52, Legal Notice No. 171 of 28 September, 2007.

Summary

Summary

Background

Sustainable water pollution control calls for effective enforcement of regulations and adoption of cleaner production technology as well as effective end-of-pipe treatment of effluents. The final effluent quality of many municipalities and industries in Kenya seldom comply with government-prescribed effluent discharge guidelines. There is, therefore, a need for a sustainable technology that can reliably achieve acceptable effluent quality for discharge into the environment at minimal cost. Natural and artificial wetland systems have been used as a cost-effective alternative to conventional wastewater treatment methods for improving final effluent quality. Data and information on pulp and paper mill wastewater treatment in constructed wetlands are few while performance data that can guide design and operation under tropical environment conditions are lacking.

This study was undertaken to explore the potential of a constructed wetland to improve the quality of the final effluent from Pan African paper mills (E.A.) Limited (PANPAPER) in western Kenya in order: 1) to be in compliance with national discharge regulations, and 2) to protect the receiving aquatic environment, the River Nzoia, and downstream riparian users. In this thesis the problematic wastewater components were characterised (Chapter 2). The data were used to evaluate the performance of the PANPAPER Mills wastewater treatment ponds and the wetland system with respect to removal of nutrients, organic matter (BOD, COD), suspended solids (TSS), and phenols (Chapters 4-6) under various operational conditions.

A pilot-scale constructed wetland covering a total area of 48.5 m^2 was located in the tree nursery just below the final stabilization pond of PANPAPER Mills. It consisted of eight subsurface flow (SSF) cells each of dimensions 3.2 m (length) $\times$ 1.2 m (width) $\times$ 0.8 m (depth) cells and two cells of dimensions 6.2 m (length) $\times$ 1.5 m (width) $\times$ 0.8 m (depth). The latter were initially operated as free water surface flow and later as subsurface flow systems. The subsurface flow cells were planted in pairs with *Cyperus immensus*, *Typha domingensis*, *Phragmites mauritianus* and *Cyperus papyrus*. The *Cyperus immensus* did not establish well due to frequent attacks by vermin monkeys and were therefore removed after eight months and the cells left unplanted. The larger cells were planted with *Typha domingensis*. All cells were filled with gravel to a depth of 30 cm.

The experimental systems operation was dynamic and ran for a total period of 3 years from 2002 to 2005. It involved different operation modes, hydraulic loading rates and retentions in order to optimise pollutant removal while maintaining good plant vitality. Initially the wetland was operated on a batch load-drain mode starting with 5-day retention time (batch phase 1). It was assumed that this format would enhance organic matter degradation. Plant vitality was relatively poor and was partly attributed to low nitrogen loading at a long retention time. A shorter retention time of 3 days (batch phase 2) was subsequently used. Although plant vitality increased, there was a reduction in treatment efficiency with respect to TSS. It was therefore decided to use a continuous flow operation mode. The results of batch phase 2 had shown better wetland performance at 3 days than at 5 days. A third phase of batch operation was undertaken as a repeat of the first phase when the wetland was considered to be mature. In all there were three phases of batch operation and two of continuous flow. In the first phase of continuous flow plant growth was at steady stage while in the second phase the plants were at an exponential growth stage. A tracer study using lithium chloride was conducted in the first phase.

Wetland performance

The study revealed that the PANPAPER Mills pond system was actually performing well as per its type and design. However, the concentration of pollutants in the final effluent (average 45±3,

394±340, 52±6, and 0.64±0.09 for BOD, COD, TSS and phenols, respectively) discharged into the Nzoia River does not comply with the national discharge limits. Mean total nitrogen and phosphorus were about 3 mg/l and 0.7 mg/l, respectively giving a low N:P ratio.

Evapotranspiration (8 -16 mm/day) was found to be an important component of outputs in the water budget of the wetland system making up to 15 – 32 %, depending on the system type. ET rates were different for the aquatic plant species studied. It was not possible to deduce the actual retention time and other hydraulic parameters (efficiency and number of "tanks in series") under continuous flow, as there was no discernable tracer concentration curve for all wetland cells. For this wastewater, which has high organic matter content, the study should be conducted with a different tracer. Alternatively, lithium chloride may still be used but with continuous feed instead of pulse feed, as was the case in this study.

Plant tissue nutrient concentrations were lower than in healthy natural wetland plants. Nitrogen concentrations based on dry weights in *Phragmites*, *Typha* and papyrus were 9.2±0.7, 7.4±0.5, and 6.1±0.2 mg/g, respectively while phosphorus concentrations were 1.7±0.12, 1.9±0.11, 1.6±0.14 mg/g, respectively. Despite this, *Typha* and *Phragmites* had satisfactory aboveground biomass production (10896 g/m^2 and 3015 g/m^2 dry weight, respectively) when compared to natural wetlands. The growth of papyrus was sub-optimal with an aerial biomass of 3075 g/m^2. In general, plant vitality and growth was lower during batch mode wetland operation. Below ground root and rhizome growth was variable. *Typha* roots penetrated the entire bed depth (approx. 30 cm) while *Phragmites* and papyrus rooting depths were in the top 20 cm and 10 cm, respectively.

Plant uptake of nutrients exceeded inputs by influent in the exponential growth stage. Nutrient mass flows indicated that in this low loaded system mineralisation and cycling of nutrients in accumulated sediments and/or in senescing/decaying plant organs are important for sustaining plant growth.

Mean removal efficiency for total nitrogen was in the range of 49 - 75 % for planted cells and 42 – 49 % for unplanted ones in continuous flow. For phosphorus removal efficiencies were 30 - 60 % in planted cells and (minus) 4 – 38 % in unplanted ones. Removal efficiency of up to 25 % may be attributed to uptake into plant shoots. Harvesting of plant shoots should be appropriately timed to avoid depletion of the nutrient pool. The removal efficiencies were lower during batch operation modes.

The constructed wetland effectively removed BOD (up to 90 %) and TSS (up to 81 %) from the wastewater to concentrations below that prescribed by the regulating authority in Kenya. However, COD removal was low (up to 52 %). The non-zero background concentration for BOD varied between 4.3 and 7.4 mg/l for the different cells while areal BOD reaction rate constants varied from 0.055 – 0.114 m/day (20 – 42 m/yr). The reaction rates are reported for pulp and paper mill wastewater for the first time. *Typha* cells had consistently higher TSS removal efficiency than *Phragmites* and papyrus in continuous flow. Besides TSS removal in the wetland bed with developed plant roots, the presence of macrophytes does not seem to enhance BOD and COD removal when compared to unplanted cells. However, the presence of plants is essential for nutrients and phenol removal.

Mean phenol removal efficiencies based on mass flows ranged from 73 % to 96 %. Good buffering was achieved even during the highest inflow phenol concentration of 1.3 mg/l and the highest hydraulic loading rate (HLR) of 9.8 cm/day in the wetland. For batch operation, optimal removal was achieved at 5-day hydraulic retention time with a mean outflow concentration of 0.053 ± 0.004 mg/l. The major processes of phenol removal are microbial breakdown (60 %) followed by sedimentation/adsorption (up to 30 %). However, ultimate biodegradation in a mature wetland may be higher as some of the sedimented and/or adsorbed phenols may be re-cycled and microbially degraded. The removal was enhanced by wetland age and presence of aquatic macrophytes especially when they were at an exponential growth stage. During this stage plant uptake rates were

5.4 – 12.7 mg phenol/day accounting for 10 – 23 % of phenol removed in SSF cells. The study reports, for the first time, phenol reduction rates (k values) in a constructed wetland. Average k, for SSF cells was 27 m/yr and 44 m/yr respectively at hydraulic loading rates of 4.1 - 4.9 cm/day and 9.8 cm/day, respectively. The reduction rate was 38 m/yr for the free water surface flow cells at a HLR of 9.3 cm/day. Mean volumetric removal rate for batch operation was 0.57 d^{-1} for *Phragmites* and papyrus cells.

Water quality of Nzoia River

PANPAPER Mills's effluent discharge in a "business-as usual" scenario causes an increase in the concentration of pollutants downstream (3 km) compared to an upstream location (500 m). The increase is highest during low flow in the River, usually in the dry season (January to March). This was reflected in an increase in the concentration of various pollutants by between 20 % and 120 % at the downstream sampling point during the month of March. From the findings of this study I predict that the quality of water in River Nzoia downstream of the discharge point would improve significantly if a full-scale constructed wetland with similar performance as that in the pilot study were to be integrated with the existing treatment ponds as a tertiary stage. Such an intervention would decrease pollution by more than 90 % for phenols. TSS, COD and BOD would be reduced by 44 %, 30 % and 63 %, respectively. Nitrogen concentration downstream of discharge would remain the same as that in the upstream location (100 % pollution reduction) while phosphorus concentrations would reduce by 50 % of the current level.

An appraisal of these findings is given in Chapter 7, which also includes recommendations for design, set-up and maintenance of a full-scale wetland.

Samenvatting

Samenvatting

Achtergrond

Voor een duurzame zuivering van afvalwater is een goede handhaving van regels nodig en dient zowel gebruik gemaakt te worden van een schone productie technologie als van een effectieve zuivering van het effluent. De uiteindelijke kwaliteit van het effluent van veel steden en industrien in Kenya voldoet zelden aan de door de overheid voorgeschreven richtlijnen voor te lozen afvalwater.

Daarom is er een duurzame technologie nodig die op een betrouwbare wijze kan zorgen voor een acceptabele kwaliteit van het effluent tegen minimale kosten. Natuurlijke en aangelegde moerassystemen zijn gebruikt als een goedkoop alternatief voor de traditionele zuivering van afvalwater.

Er is nauwelijks informatie over de zuivering van het afvalwater van papierfabrieken door middel van kunstmatige zuiveringsmoerassen, terwijl gegevens die kunnen bijdragen aan een goed ontwerp geheel ontbreken.

Dit onderzoek is uitgevoerd om de mogelijkheid van het gebruik van een helofyten filter voor verbetering van de kwaliteit van het uiteindelijke effluent van de in het westen van Kenya gelegen papierfabriek van Pan African (E.A.) Ltd. te onderzoeken, teneinde te voldoen aan de nationale lozingseisen, en om het ontvangende aquatische milieu, de rivier Nzoia, en de benedenstroomse gebruikers te beschermen. In hoofdstuk 2 van dit proefschrift zijn de verschillende componenten van het afvalwater bepaald die de meeste problemen geven. De gegevens zijn gebruikt om de efficientie van de stabilisatievijvers van de PANPAPER fabriek en van de helofytenfilters onder verschillende operationele omstandigheden te bepalen voor wat betreft de verwijdering van nutrienten, organisch materiaal, (BZV, CZV), gesuspendeerd materiaal en fenolen (Hoofdstukken 4 – 6).

Een experimenteel helofytenfilter met een totaal oppervlak van 48,5 m^2 werd aangelegd in de boomkwekerij net achter de laatste stabilisatie vijver van PANPAPER. Dit helofytenfilter bestond uit acht wortelzone compartimenten elk 3,2 m lang en 1,2 m breed en 0,8 m diep en twee compartimenten met afmetingen van 6,2 m x 1,5 m x 0,8 m. De laatste 2 compartimenten werden oorspronkelijk gebruikt als vloeiveld compartimenten en later als wortelzone compartimenten. De wotelzone compartimenten werden twee aan twee beplant met *Cyperus immensus*, *Typha domingensis*, *Phragmites mauritianus* en *Cyperus papyrus*. De *Cyperus immensus* gedijde niet erg goed, omdat de stengels opgegeten werden door apen. Daarom werden de planten na 8 maanden verwijderd en werden deze compartimenten verder niet meer beplant. De grote compartimenten werden beplant met *Typha domingensis*. Alle compartimenten werden gevuld met een 30 cm dikke kiezellaag.

De helofyten filters waren gedurende 3 jaar (van 2002 tot 2005) operationeel. De proefomstandigheden, zoals hydraulische belasting en verblijftijd, werden steeds aangepast, teneinde de nutrienten verwijdering te optimaliseren en de planten vitaal te houden. In het begin werd het helofyten filter elke 5 dagen geleegd en weer opnieuw gevuld (batch fase 1). Er werd verondersteld dat deze methode de afbraak van organisch materiaal zou verhogen. De vitaliteit van de planten was relatief slecht en dit was ten dele toe te schrijven aan de lage stikstof conentraties in het water en de lange verblijftijd. Vervolgens werd een kortere verblijftijd van drie dagen gebruikt (batch fase 2). Hoewel de planten er gezonder uitzagen was er een verslechtering van de verwijdering van totale hoeveelheid zwevend materiaal. Daarom werd besloten om over te gaan tot een continu cultuur. De resultaten van batch fase 2 met 3 dagen retentietijd waren beter dan bij 5 dagen. Een derde batch fase werd gebruikt om deze fase te vergelijken met de eerste fase toen het filter nog jong was. In totaal waren er drie batch fases en twee met continue doorstroming. In de eerste fase van continue doorstroming waren de planten in een stationaire groeifase, terwijl ze in de tweede fase exponentieel groeiden. In de eerste fase werd ook nog een tracer studie uitgevoerd met lithium chloride.

Functioneren van het zuiveringsfilter

Dit onderzoek laat zien dat de zuiveringsvijvers van de PANPAPER papierfabriek eigenlijk goed voldeden aan de ontwerpeisen. Echter, de concentraties van verschillende stoffen in het effluent (gemiddeld 45±3, 394±340, 52±6, en 0,64±0.09, respectievelijk voor BZV, CZV, TZS en fenolen) voldoen niet aan de nationale lozingseisen. Het gemiddelde totaal stikstof en fosfor gehalte was respectievelijk 3mg/l en 0,7 mg/l, resulterend in een lage verhouding tussen N en P.

De verdamping door de planten (8-16 mm per dag) bleek een belangrijke component van de waterbalans, en bedroeg, afhankelijk van het type compartiment, 15 - 32 %. Het bleek niet mogelijk om de echte retentietijd en andere hydraulische parameters te bepalen voor het continue doorstroom systeem, omdat er geen goede tracer concentratie grafiek gemaakt kon worden. Voor dit soort afvalwater met een hoog organische gehalte zou eigenlijk een andere tracer gebruikt moeten worden. Lithium chloride zou eventueel nog wel gebruikt kunnen worden als het continu toegediend wordt in plaats van eenmalig, zoals hier het geval was.

De concentraties van voedingsstoffen in de weefsels van de planten was lager dan in gezonde moerasplanten in de natuur. De concentraties stikstof bepaald als drooggewicht in Phragmites, Typha en papyrus waren respectievelijk 9,2±0,7, 7,4±0.5 en 6,1±0.2 mg/g, terwijl de fosfor concentraties respectievelijk 1,7±0,12, 1,9±0,11 en 1,6±0,14 mg/g bedroeg. Ondanks dit hadden Typha enPhragmites vergeleken met natuurlijke moerassen een behoorlijke productie van bovengrondse biomassa van respectievelijk 10.896 g/m2 en 3.015 g/m2. De groei van papyrus was niet helemaal optimaal met een bovengrondse biomassa van 3.075 g/m2. In het algemeen waren de planten tijdens de batch fase minder gezond.

Er was duidelijk verschil in de wortels: de wortels van Typha groeiden diep (ongeveer 30 cm), terwijl Phragmites en papyrus slechts tot 20 en 10 cm diepte groeiden.

De opname van nutrienten door de planten was groter dan de aanvoer in het influent in de exponentiele groeifase. De nutrienten massa balans gaf aan dat in deze laag belaste systemen mineralisatie en het hergebruik van nutrienten uit afgevallen bladeren en stengels belangrijk is voor de groei van de planten.

De gemiddelde verwijdering voor totaal stikstof bedroeg 49 – 75 % voor beplante compartimenten en 42 – 49 % voor onbeplante compartimenten onder continue doorstroming. De verwijdering van fosfaat bedroeg 30 - 60 % in beplante compartimenten en min 4 (!) – 38 % voor niet beplante compartimenten. Ongeveer 25 % kan waarschijnlijk toegeschreven worden aan de opname door de stengels van de planten. Het tijdstip voor de verwijdering van plantenstengels moet zorgvuldig bepaald worden teneinde een tekort aan nutrienten te voorkomen. De efficientie van de verwijdering was lager bij de batch systemen.

Het helofyten filter was in staat om BZV (tot 90 %) en TZS (tot 81 %) uit het afvalwater te verwijderen tot concentraties onder de lozingseisen van de Kenyaanse overheid. De CZV verwijdering was echter laag (tot 52 %). De achtergrond concentratie van BZV lag tussen 4.3 en 7.4 mg/l voor de verschillende compartimenten terwijl de eerste-orde BZV (reactie) constante varieerde van 0,055 – 0,114 m/dag (20 – 42 m/jaar). Dit is de eerste keer dat deze reactie snelheden gegeven worden voor afvalwater van een papierfabriek. Tijdens een continue doorstroming hadden compartimenten beplant met *Typha* altijd een hogere verwijdering van TZS dan *Phragmites* en papyrus. Uitgezonderd de verwijdering van TZS in een compartiment met goed ontwikkelde plantenwortels schijnt de aanwezigheid van planten (in vergelijking met onbeplante compartimenten) de verwijdering van BZV en CZV niet te beinvloeden. De aanwezigheid van planten is echter essentieel voor de verwijdering van nutrienten en fenol.

De gemiddelde verwijdering van fenol gebaseerd op massabalansen lag tussen 73 en 96 %. Een goede verwijdering werd zelfs verkregen bij de hoogste fenol concentratie van 1.3 mg/l en bij de hoogste hydraulische belasting van 9.8 cm/dag. In het batch systeem werd de hoogste verwijdering verkregen bij een hydraulische retentietijd van 5 dagen met een gemiddelde effluent concentratie van 0,053 ± 0,004 mg/l. De belangrijkste verwijderingsprocessen voor fenol zijn microbiele afbraak (60 %) gevolgd door sedimentatie/adsorptie (tot 30 %). Echter de uiteindelijke biodegratie kan hoger uitpakken doordat gesedimenteerde of geadsorbeerd fenol weer vrijkomt en alsnog door bacterien wordt afgebroken. De verwijdering was hoger naarmate het helofytenfilter ouder was en door de aanwezigheid van waterplanten, in het bijzonder in de exponentiele groeifase. In deze groeifase bedroeg de opname 5,4 – 12,7 mg fenol/dag resulterend in 10 – 23 % van het fenol verwijderd in wortelzone compartimenten. Deze studie geeft voor het eerst fenol reductie snelheden (k waarden) in een helofytenfilter. De gemiddelde k waarde voor wortelzone compartimenten was 27 m/jaar en 44 m/jaar respectievelijk voor een hydraulische belasting van 4,1 – 4,9 cm/dag. De reductie snelheid was 38 m/jaar for vloeiveld compartimenten bij een hydraulische belasting van 9,3 cm/dag. De gemiddelde volumetrische verwijderingssnelheid tijdens batch bedroeg 0,57 dag^{-1} voor met *Phragmites* en papyrus beplante compartimenten.

De waterkwaliteit van de Nzoia rivier

Indien de PANPAPER papierfabriek op de gebruikelijke wijze het afvalwater zuivert in de zuiveringsvijvers dan veroorzaakt dit vergeleken met het inname punt stroomopwaarts (500 m) een toename van vervuilende stoffen stroomafwaarts (3 km). Deze toename is het hoogst gedurende lage waterstanden in de rivier in de droge tijd (januari tot maart). Dit was te zien aan een toename van de concentratie van verschillende stoffen van 20 tot 120 % bij het benedenstroomse meetpunt in maart. Uit de bevindingen in dit proefschrift kan voorspeld worden dat de waterkwaliteit in de rivier Nzoia beduidend zal verbeteren als er een helofytenfilter aangelegd zou worden dat groot genoeg is om het water uit de bestaande bassins na te zuiveren. Het helofytenfilter zou dan de derde stap kunnen zijn. Zo'n interventie zou de concentratie fenolen in de rivier met 90 % verlagen. TZS, CZV en BZV zouden verlaagd worden met respectievelijk 44 %, 30 % en 63 %. De stikstof concentratie zal stroomafwaarts hetzelfde zijn als bovenstrooms, terwijl fosfaat waarden 50 % lager zullen zijn.

Een overzicht van deze gegevens kan gevonden worden in hoofdstuk 7, en daar kunnen ook aanbevelingen gevonden worden voor het ontwerp, opstarten en onderhoud van een grootschalig helofyten filter.

Author's Resumé

Margaret Akinyi Abira was born on March 18, 1959 in Nairobi, Kenya where she attended her primary and secondary education. After completing her higher secondary education at the Alliance Girls' High school, she proceeded to Nairobi University in 1979 where she undertook a Bachelor of Science degree course majoring in Chemistry. She was employed by the then Ministry of Water Development as a trainee Chemist in 1983. Margaret rose through the ranks serving in various capacities including heading the Central Water Testing Laboratory until 1992 when she obtained a scholarship from the Netherlands Fellowship Program to study Water Quality Management at UNESCO-IHE in Delft, the Netherlands. She graduated with a postgraduate diploma in 1993 and a M.Sc. degree in 1994.

On returning to Kenya, Margaret was appointed to head the National Water Quality Monitoring Program. She participated in various national programs including the National Environmental Impact Assessment (EIA) Program, the drafting of the EIA guidelines and procedure and the UNEP/UNDP project on the development and harmonization of environmental standards in Kenya. She was the representative of the water sector in the Inter-Ministerial Committee on Environment and the steering committee of the Biodiversity Data Management Project in Kenya between 1996 and 1998.

She joined the Lake Victoria Environment Management Project (LVEMP) in 1998 as the Task Leader of the subprogram on the Management of Industrial and Municipal Effluents. In January 1999 she commenced her PhD study at UNESCO-IHE Institute for Water in a sandwich program. The same year she became the Team Leader of the Integrated Tertiary Industrial Effluents pilot project of the LVEMP under which she undertook the present research study.

Margaret was appointed to the position of Chief Chemist in May 2004. Later in November she became the Regional Manager of the Lake Victoria South Catchment Area under the newly established Water Resources Management Authority (WRMA). She held the post until July 2005 when she proceeded to Delft to write her thesis. She is currently the Operations Manager of WRMA.

Margaret's research interests include: wetlands systems – ecotechnologies for water management, water resources and environmental management, climate change issues, and gender and environment issues.

Publications

International journals with peer review

Abira, M.A., van Bruggen, J.J.A. and Denny, P. 2005. Potential of a tropical subsurface constructed wetland to remove phenol from pre-treated pulp and papermill wastewater. *Water Science and Technology* 51 (9): 173 - 176.

Abira M. A, Ngirigacha H. W. and van Bruggen J.J.A. 2003. Preliminary investigation of the potential of four tropical emergent macrophytes for treatment of pre-treated pulp and papermill wastewater in Kenya. *Water Science and Technology* 48 (5): 223-231.

International proceedings with peer review

Abira, M.A., J.J.A. van Bruggen, P. Denny, 2004. Potential of a tropical subsurface constructed wetland to remove phenol from pre-treated pulp and papermill wastewater Proceedings of the 9th International Conference on Wetland Systems for Water Pollution Control, September 2004, Avignon, France.

Abira, M.A., H.W. Ngirigacha and J.J.A. van Bruggen, 2002. Preliminary Investigation of the Potential of *Cyperus immensus, Cyperus papyrus, Typha domingensis,* and *Phragmites mauritianus* for Treatment of Pre-Treated Pulp and Paper Mill Wastewater in Kenya, Proceedings of the 8th International Conference on Wetland Systems for Water Pollution Control, September 2002, Arusha, Tanzania, pp.514-525.

Books with peer review

Ogola, J.S., M.A. **Abira** and V.O. Awuor (editors) 1997. Potential Impacts of Climate Change in Kenya, Climate Network Africa Publ. ISBN 9966-9949-0-4.

Abira, M.A., C.O. Oleko, J.O. Okungu, J. Abuodha, and R.E. Hecky, 2005. Industrial and Municipal Effluents Loading into Lake Victoria Catchment, Kenya. National Water Quality Synthesis Report Kenya. Lake Victoria Environmental Management Project Report on Water Quality and Ecosystem Status in Kenya, pp 112-138.

Abira, M.A., 1997. Water Resources. In Ogola, J.S., M.A. Abira and V.O. Awuor (eds.). Potential Impacts of Climate Change in Kenya, Climate Network Africa Publ. ISBN 9966-9949-0-4, pp. 43-53.

Onyango, J.C.O.; E.O. Massawa and M.A. **Abira**, 1997. Biodiversity and Climate Change. In Ogola, J.S., M.A. Abira and V.O. Awuor (eds.). Potential Impacts of Climate Change in Kenya, Climate Network Africa Publ. ISBN 9966-9949-0-4, pp. 55-62.

Abira, M.A., 1997. Climate Change and Human Health. In Ogola, J.S., M.A. Abira and V.O. Awuor (eds.). Potential Impacts of Climate Change in Kenya, Climate Network Africa Publ. ISBN 9966-9949-0-4, pp. 151-158.

Ogola, J.S. and M.A. **Abira**, 1997. Responses to Climate Change. In Ogola, J.S., M.A. Abira and V.O. Awuor (eds.). Potential Impacts of Climate Change in Kenya Climate Network Africa Publ. ISBN 9966-9949-0-4, pp. 169-178.

Others

Kirui, M.L. and M.A. **Abira**, 1996. Environmental Impact Assessment for Biodiversity Conservation, in Bk: Institutional Support for the Protection of East African Biodiversity, Proceedings of the Kenya Workshop on Biodiversity, UNO/RAF/006/GEF Field Document No. 19.

Abira, M.A., 1995. Climate Change and Wetlands: Expected Impacts. Climate Network Africa Newsletter Impact No. 16, 24 pp.

Abira, M.A., 1994. Nutrient Uptake by Periphyton Communities. M.Sc. Thesis E.E. 140, International Institute for Infrastructural, Hydraulic and Environmental Engineering (IHE), Delft, The Netherlands.

T - #0100 - 071024 - C26 - 254/178/10 - PB - 9780415467155 - Gloss Lamination